R. E Fullerton

An American Italy for Invalids

A dissertation, showing the advantages, incidents, etc., of a journey on the

plains, in the Rocky Mountains and Mexico

R. E Fullerton

An American Italy for Invalids
*A dissertation, showing the advantages, incidents, etc., of a journey on the plains,
in the Rocky Mountains and Mexico*

ISBN/EAN: 9783337127930

Printed in Europe, USA, Canada, Australia, Japan

Cover: Foto ©berggeist007 / pixelio.de

More available books at **www.hansebooks.com**

A DISSERTATION,

SHOWING THE

ADVANTAGES, INCIDENTS, ETC., OF A JOURNEY ON THE

PLAINS, IN THE ROCKY MOUNTAINS

AND MEXICO,

FOR THE CURE OF ALL

Vol. 1.   No. 1.

ksellers,

o.

OUR GREETING.

We come with a cheering offering to the subject of dark despair. We come with a new-born hope to those before whom the grim monster, Death, would wave his icy pall. We come to dispel, as it were, a fearful sirocco, not by imaginary and shadowy theories, but by *facts already established.* We come, not with vague and unintelligible hypotheses, based upon idle conjecture; nor with narcotic nostrums which intoxicate, please, and ultimately kill, but with nature's sanatives, emanating from natural sources, under the influence of natural laws. We come, not with glowing descriptions of fertile valleys, exhaustless pasturage, cargoes of golden nuggets, great emporiums of Mammon's glittering livery, but with that which has no measure of value in gold or gems or precious stones—the means of LIFE,—life in its highest meaning, wherein every sense shall have its full development, and joy in its own activity. We would come in that spirit which animated a nation's heart with pulsations of divinest sympathy for the poor raftsman who, temporarily protected by rocky obstructions, was at length, in spite of all efforts to save him, finally engulfed in the raging torrent of Niagara. Business, everywhere, was suspended; all stood appalled, awaiting his sad fate, little dreaming that around them were thousands equally though not so speedily doomed; unnoticed, because the abrasions of insidious disease upon the brittle thread of life are less marked in their character than these sudden catastrophes. And when our shortcoming shall have shown that only master-minds should have touched upon our theme, we shall still feel that we have endeavored to discharge a duty in Humanity's cause.

4

The Italy of America.

Chance, more than scientific research on so important a subject—a subject replete with life or death to scores of thousands—seems at first to have discovered that the climate of our western territories and Mexico is unequalled in salubrity by that of any other country. The huntsman, the trapper, who had drank too deeply the malaria of the lower Mississippi, Missouri, Arkansas, and other sluggish and miasmatic streams; who, by a life of indiscretion and irregularity, in swamps where every drop of water was putrid and every breath of air charged with a cause of disease and death—"marsh miasma,"—was worn down by chronic intermittents, his eyes jaundiced, skin swarthy, liver torpid, spleen enlarged, kidneys affected, blood watery, muscles soft and flabby, ankles swollen; until he felt the ravages of slow but unmistakably fatal disease undermining the mainsprings of life; no longer able to cope with the duties of his vocation, and impressed with a mortal horror of early death, joined the first caravan westward bound with some hope of relief from the change. He may now be found here and there along the frontiers of civilization, the "live man" of more than three score and ten years, having reared a mixed-blood Indian or Mexican family, yet active and useful among the people of his adoption; whereas no rational person can believe that he could have lived long where and as first described. The debauchee of city life, exhausted by excessive dissipation until bankruptcy of purse and person supervened, looked to the mines of these regions to conceal his disgrace and replenish his impoverished exchequer. Soon the invigorating atmosphere of these usually cloudless skies gave him a physical power which stimulated his mental energy; now he looms up as a tower of strength, lucratively working mining machinery. And so of the deserter soldier; his constitution shaken by chronic chills from being stationed at badly-located posts until his blood was so thinned that it scarcely stimulated his brain, and caused a lethargic torpor which prevented response to the *reveille*. To avoid punishment he fled to the enemy it was his duty to confront. Now his stalwart form is met in the van of the most efficient teamsters of the plains. Instances of great improvement of health, very remarkable recoveries from consumption and the most inveterate chronic diseases having become very frequent, train runners were in the habit of proposing to invalid friends a trip on the plains, laughingly telling them it should cost nothing unless they entirely recovered and gained a large number of pounds of flesh, in which event ten dollars per pound would be charged on the weight gained.

Learning these happy results, many invalids, during the last few years, have availed themselves of these climates; some as tourists, others permanently engaging in some business of the country, and, so far as can be ascertained, *invariably* with great improvement of health, most of them entirely recovering, justifying us in the opinion that under the more favorable auspices of well-conducted and more extended journeys, reaching across greater territory and migrating with the seasons from points far north in the Rocky mountains to points south in Mexico, even more encouraging results may be reasonably expected and certainly realized.

Of persons thus cured hundreds may be found in traveling through the prairie and mountainous parts of the Indian nations, Texas, Mexico, New Mexico, and the western territories. Many of these were, ten to fifteen and eighteen years ago, reduced to mere living skeletons by hemorrhage of the lungs, heavy expectorations, dyspepsia and other forms of chronic disease. The history of their sickness, when compared with their excellent health now, seems incredible; but I, being similarly afflicted, was particular to get *facts*, not extravagant exaggerations. I have the privilege of using the names of a few of these persons, and others I shall, with no impropriety, advert to. I shall rely on memory for statements given me by them.

About seventeen years ago HENRY C. KING, ESQ. (then of Georgia, and nephew of the distinguished statesman of the same name,) was pronounced by his physicians consumptive. Being of a consumptive habit, (all others of his father's family, I believe, have died of that disease,) every reputable means were tried with a view to a cure. Growing worse, he determined to take the then hazardous chances of a life on the plains and among the Rocky Mountains, and made many journeys to these now favorite resorts. Hemorrhage, evening fevers, night sweats and heavy expectorations had at one time reduced him so low that a gentleman told me he was in daily expectation of being summoned to his funeral. He now is an active member of the bar of western Texas. He says that trips across the plains saved his life, and that if he had a relative given up by physicians he would start with him in an ambulance to these countries. To this last sentence *I owe my life.* Until *moved* by it I never fully recovered. Never shall I forget the affectionate tone in which he uttered to me the following: "If my dear brother had gone with me to Pike's Peak and accompanied me at the approach of winter through New Mexico he now, quite probably, would be living."

MR. JOHN R. BAILEY, a very intelligent saddler at San Antonio, now in perfect health, says that he was reduced very low by all the symptoms of consumption; was regarded as hopeless, but was perfectly restored by journeying across the plains. His whole manner and appearance is that of very vigorous health. He speaks of the disease and its successful management as a very light affair, the only remedy necessary and the only inconvenience incidental to its cure being that of camp life on the plains. Although very prosperously engaged in the harness trade, he says he will close out and again start if any symptom of the disease reappears.

A merchant of Procedia Delnorte, about twenty years ago, was as low as either of the gentlemen above referred to. Although now quite an old man he is very healthy, fleshy, active and jovial. Traveling through and living in this country cured him. He only visits occasionally his business house in town, which is on the Rio Grande—will not spend a night there, but returns to his ranch in the hills, seventeen miles off. He told me I need go no further than home with him. He had a land claim in view for my selection, on which he would guarantee I could live until the Apaches or Comanches would scalp me! I accepted his wit and kindness, but declined the startling invitation.

Mr. H——, a man nearly eighty years old, about three or four years ago went to the Rocky Mountains extremely low. It appeared that he could not live more than a few weeks. He, last summer, did full labor on his farm, near Colorado City, and is now quite an active man.

Dr. Southerland, eminent and now actively useful in the dental profession at Colorado City, went to Colorado last summer a mere living skeleton —so low of consumption that a letter was received with instruction if he was dead to return it to certain parties. In about three months this gentleman gained sixty-three pounds of flesh!

Hundreds of equally remarkable recoveries could be given, but I will now speak of myself and thus close this part of the subject.

Early in life I had occasion to fear consumption. I decided to acquire a knowledge of the medical profession as a means of health. As that disease then appeared to be incurable by remedies of the regular profession, I sought for Indian and other articles not used by physicians of the old school. Eager for proficiency in that direction, I was easily seduced by the popular " isms " of the time, and gave them all a careful and thorough examination. Even that system which would, by trituration, agitation, shakings, &c., increase the power of the millionth part of a grain of opium until it would have sufficient potency to give all the inhabitants of the earth their last sleep, was fully canvassed.

I received thorough instruction in the Botanic and Eclectic systems of medical practice. Finding that the term "eclectic," as applied to medicine, should mean that system which chooses from the mineral, vegetable and animal kingdoms those remedies which experience shows to be the most safe and successful, and knowing that the allopathic was the only school of these vast resources, I graduated in the Medical College of Ohio; afterwards matriculated at New Orleans; after a full course there took tickets to Bellevue and New York hospitals, in order to learn the disease as found in different parts of the country.

Although greatly exposed to every vicissitude of weather, in a full practice for over twenty years, yet, by cautious habits, very little medicine, and with occasional inhalations to check catarrhal attacks, I prevented any lesion of pulmonary tissue until four years ago, when great exposure and very bad digestion appeared to develope the disease in frightful form. Very copious expectorations, spitting of blood, evening fevers, night sweats, indigestion, &c., reduced me very low. Dr. Bruns, of New Orleans, eminent in that specialty of his profession, declared that he found extensive tuberculous deposits in both lungs, and softening in the left! There was no time to lose! In a few hours I was on my way to that climate to which, thanks to a kind Providence, I owe my life. My health is now excellent—have gained from 118 to 157 lbs. As an evidence of my entire recovery, and at the risk of appearing *homely* in my statements, I must here say that I had no difficulty in getting a policy from a life insurance company after giving the medical examiner all these facts.

I left my son at school in New Orleans. I hoped he would outgrow a predisposition to lung disease. In that heavy, humid and vitiated atmosphere, a light, hacking cough changed to alarming symptoms. Hemorrhage wet through a mattress. Dr. Richie and other physicians feared he would pass into "galloping consumption." As soon as able he joined me. Now, thanks to the Giver of every good and perfect gift, I can challenge the whole country for a healthier boy. As well attempt to weave Heaven's light into a web of darkness as to show that this is not the climate for the cure of such disease.

Migratory Journeying Proposed.

If an inordinate ambition to effect an object is frailty, then have we shown weakness by an intense desire to establish beyond all possibility of doubt that by migratory life in suitable climates, diseases which cannot be removed by any other known means, *can be cured.* Fortunately for and corroborative of this opinion, is an established usage which has borne the test of centuries, and worn better than any other course pursued for the cure of chronic ills, *i. e.,* a sojourn away from home. All physicians recommend and all intelligent persons recognize the advantages of a trip *somewhere* as a preventive of debility, a restorative in convalescence, and a curative of protracted indisposition. New Orleanians go north or east during summer. At their points of destination they find the inhabitants arranging jaunts north—further north, and the advantages of change are alike to all.

My proposition is to meet those who will accompany us at Omaha, on the Union Pacific Railroad; visit and spend the summer in mountain parks west of a line drawn from Cheyenne to Pueblo, near the source of the Arkansas river; by fall reach New Mexico, travel far south through the Republic of Mexico next winter, and return in the spring.

Our line of travel lies southward of Cheyenne, the total distance about two thousand miles, through the most interesting parts of these territories and Mexico. See further description elsewhere.

Advantages of Migratory Living.

We shall attempt a consideration of this momentous subject, under the following heads, in the order here named: Necessity of Changes; Climate; Digestion; Mental Diversion; Exercise; Altitude; Regularity of Habits; Remedies, &c.; Curability of Consumption; Preparation for the trip;

NECESSITY OF CHANGE, TOTAL CHANGE, AND FREQUENT CHANGE FOR THE CURE OF CHRONIC DISEASE.

A law of our nature is that our organism is so constituted as to become inured to influences, internal and external, until tolerance is established of excitants which at first would be fatal. In obedience to this law all living things, animal and vegetable, adapt themselves to necessity shaped by their surroundings.

Of this all persons have many illustrations. " Fire Kings " have so accustomed themselves to intense heat that at exhibitions they expose themselves to temperature which would kill those unused to it, whilst north polar explorers achieve the reverse feats. Such tolerance may be of the whole body or any part of it. Those who work at forges handle metals which quickly would lift the flesh off the hands of ice venders. A Mexican with bare feet, on the frozen ground, his face covered with a *serape*, was asked by a heavily-booted but barefaced man if his feet were cold, which he answered by querying, " Is your face cold?" During the recent unpleasantness, Virginia soldiers lay on one half of a lean blanket, covered with the other half, on frozen ground; and yet being used to it, they declared they slept comfortably.

This adaptation to changed circumstances can be traced to the shaping of animals and vegetation. It is plausibly said gay appearances are given growing horses by high troughs.

After adaptation to the new hygiene, the new excitant fails to have a strong effect. A change of stimulus would be necessary to another commotion. Take subjects of narrow chests and contracted lungs from low altitudes, where the dense atmosphere furnishes, in small volumes, all the oxygen necessary for respiration, to an altitude of ten or fifteen thousand feet, where a larger volume of the rarified atmosphere is requisite for respiration. Hurried and laborious efforts at breathing will continue for many weeks until the chest and organs of it are enlarged to their necessity, after which man and beast breathe naturally. Or take the slender oak from a dense forest in a low valley where a strong breeze seldom stirs, and therefore where there is no use of a strong stem; transplant it to a high and exposed place; after shaped to the change there will be the sturdy tree of large shaft, thick, strong limbs, and deep and wide roots.

More closely to approach our subject we will show that a keen susceptibility of our race to new stimulants exists, and that by *continued use* many excitants, in a great measure, lose their power. This is true of many drugs. The anti-periodical virtue of quinine against intermittent disease is almost lost by continued use, so that for chills and fever, &c., it becomes an uncertain medicine. As an anti-paroxysmal remedy, injudiciously used against disease. I have known many females whose health, as such, was ruined by quinine; and yet in after years, when the susceptibility to its strong impression was in a great measure blunted by frequent use, this medicine did not so much interfere with their periodicity as at first. The habitual opium eater can take ten times the quantity which at first would kill him. The substance of five grains of coffee or tea will keep awake a whole night persons unaccustomed to the evening use of these beverages, but ultimately they will sleep soundly under the effects of many hundred grains of either; and so of the use of whisky, tobacco, pills for constipation, and many other habits. The process of so-called seasoning or acclimatizing in unhealthy countries to an atmosphere at first sure to cause many attacks of disease, shows the same thing at a frightful cost.

Then, as susceptibility to strong impressions, whether salutary or the reverse, is ultimately lost, to wholly revolutionize a morbid train of diseased action, which gradually extinguishes vitality, we need not only a total change of climate, but, as far as possible, of all other agents which bear any relation to our necessities. It is needless to say that the change should be to a *better climate*, although it is positively known that *any* and *all changes*, even to a *worse atmosphere*, have often been attended with very happy results. This was because of the *new stimulus.* Digestion, (as we shall show under that heading,) or a vigorous assimilation of suitable nutrition, is inseparably connected with restoration from chronic disease. Without this, all is folly—hope is foundationless—all is lost! We say this from personal experience, from observation, from *demonstrable reasoning.* Now, all waters that pass through or over the earth are more or less charged with minerals, earths, gases, and various salts. These, when new to the subject of their use, excite to more vigorous action the liver in common with all the glands which furnish fluids for digestion, thus giving to tourists everywhere that good appetite and vigorous digestion which is indispensable to the cure of any chronic disease. The advantages from change of waters have been greatly underestimated. improvements of appetite and health being attributed to atmospheric change more than to the real cause.

In order to continue this commencement of cure it is necessary often to change for waters containing other medicinal ingredients, or the same substances in different combinations. It is true that any change sometimes is too exciting, showing the excessive effect by irritation of the stomach and bowels. This can be controlled by mild means and by the temporary use of distilled water, for the preparation of which a still will be taken in the expedition.

Nor are waters, air, exercise, mental diversion, &c., the only *changes* which will be enjoyed. Articles of food indigenous and produced on soils so different chemically from those of the states, so far as possible, will be procured for subsistence.

It is believed by scientific investigators that "cereals grown on the alkaline soils of the territories and Mexico are unusually rich in some of those elements, including the phosphates, now so highly extolled by physiologists in the treatment of tubercular and other diseases involving a lowered condition of the vital functions."

Because of the *new* stimulus, and not from any peculiar ingredients, all waters resorted to get reputation for curative virtues. Any water containing the same quantity of new excitants, provided they are not specifically poisonous, merits the patronage of the most celebrated watering places. And if any particular water could effect the immense results of a frequent change of waters, assisted by all the auxiliaries incidental to making changes, the combination of all knowledge on the subject, and of all earthly means, could not equal the celebrity of that water.

This, we know, will clash with the opinions of those who have had a lifetime schooling in impressions that certain waters are possessed of extraordinary powers. All that is necessary to establish the contrary is a popularization of the theory that *change, frequent change*, will effect far more than has ever been anticipated from any *one* influence.

How absurd is the opinion that, because of chemical action within the earth, certain waters, as those of the Salina in Bavaria, give a gaseous effervesence, those waters are possessed of extraordinary, ah! of supernatural power. Why, any chemist can give to any common water a similar gaseous appearance. Such opinions are in keeping with the superstitions of savages who cast sacrifices upon those waters. Cultivated minds look to other influences than *mere appearance* for healing virtues.

A visit to, a local sojourn at, watering places, is daily gaining favor, and will be practiced by increasing numbers, and followed by very fortunate results, because of the *change*. Ages will yet be required fully to show the advantages of a wider field of change. When we find the most intelligent people of our generation entertaining such opinions as the following, we should be prepared to expect mysteriously unintelligible things of them relative to healing agencies: If an infinitesimal globule of sugar of milk, charged with the one-millionth of a grain of arsenic, had fallen over St. Anthony's falls a thousand years ago, it would, by dilution, diffusion and the motion of that ever rolling volume of unmeasured fluid, have been increased in potency, and communicated to every drop of water of the Mississippi river in such power that a thousand years from now one drop taken from the mouth of that river, and merely smelled at once a month, would be sufficient to cure very susceptible subjects of the most inveterate diseases. More than this, all life, whether of man or beast, fish, fowl or forest, along that stream, as far as the winds wafted the medicated vapors arising from it, would perish. It has been shown that fumes of arsenic extinguish all vitality, animal and vegetable. This is not an extreme view of that system of resublimed nonsense called homœpathy.

By traveling, invalids avail themselves of a thousandfold the advantages of stopping at any one place, and that without increase of expense. The monotonous sameness of any place, aside from all other considerations, is very depressing to a nervous system unstrung by protracted disease.

The natural solutions of medicinal substances found in waters generally stimulate the digestive organs more naturally, mildly and effectually than artificial preparations.

It is exceedingly difficult to disabuse the minds of those who have known extraordinary results from a sojourn at any particular watering place. They believe that no other water or place or means could have effected such important results, when in fact thousands who traveled elsewhere and never saw that identical water or place, have realized as remarkable recoveries.

Major LOGAN of Denver was, long years ago, by radical change of climate, and the better atmosphere there inhaled, changed from exceedingly bad to vigorous health. Last summer he lapsed into general debility, and a disease came upon him from which he never would have recovered where his susceptibility to strong impressions was lost. I advised a journey. He went to springs near Sante Fe. In a few months he returned, so fully recovered and so much increased in flesh and improved in appearance, that acquaintances found a second look necessary to his recognition. To a remark, "Major, you are fifteen years younger than when you left," he replied, in a masculine voice and manner, quite unlike his effeminate look and tone on leaving, "Fifteen? I say forty!" The Major believes that those identical waters did the work, when if he had gone as far in any other direction, in this fine climate, the result would have been the same.

Two persons of the same sex, temperament, age and condition, and of the same disease, may go on a tour each in an opposite direction, and to waters as far the opposite of each other as may be, in chemical constituents, and yet each one will be alike improved, provided that the atmospheric and all other conditions are equally new and favorable.

Many have very greatly improved by new influences, and could have recovered as well had they continued to make necessary changes. This has been shown in numerous instances, in none better than that of the lamented Dr. EDWARD T. NORMAN, of New Orleans. Finding tubercules in his lungs, followed by hemorrhage and a rapid advance of the disease, he went to western Texas in the fall of 1869, I found him very low spirited. He spoke of the folly of a wish to live without physical vigor sufficient to make himself useful. But new influences were operating on him, notwithstanding the untoward bent of his skepticism against them ; soon his improvement was very great. He was a greatly changed man in body and mind. Lively, cheerful, hopeful, *useful*. Took an office—was a competent guardian of the public physical weal. A staunch stake holder when life trembled in the pulse. But, alas! apartments chosen for business by day and lodging at night were badly selected. In a wide, flat and fertile valley of a river, and in the center of Boerne, the streets of which town are nightly filled with herds of cattle, the whole earth and air charged with noxious effluvia, his rooms were surrounded by corn, weeds and other tall vegetation, stagnating the salubrious breezes from the surrounding hills. Frequent summer and fall rains fell—sensibility to that strong stench was lost by constant respiration by the good citizens of that place. But in passing through from my camp on an elevation beyond, I found it, when the winds were quiet, very offensive. I told the Doctor I did not like his location. He said he could do no better. Of an age at which the disease is apt to make fearfully rapid strides, his system charged with malaria, new influences lost, he fast declined. The cause of bilious attacks in that latitude during fall stubbornly locked his liver and digestive secretions. He said he could not digest enough to keep a bird alive. Inveterate indigestion, which invariably affects mental energy, in spite of his better judgment, prostrated all hope of improvement, even from change, although that fact must have been strongly established on his mind from personal experience, showing how fatally incapable the wretched dyspeptic is of knowing his vital necessities, although learned in these things. He was not himself at all. Had only one influence unfavorably affected him he quite probably would have beaten it, but he was beset and overwhelmed by a combination which unnerved him. Oh, that I could reverse the tide of time. I would importune even more strongly one so noble, one so learned as my departed friend to avail himself of another change. As it now is too late, I can only repeat his parting words : " Fare thee well." Even then, had he gone with me to Mexico, in a few days or a very few weeks at farthest he quite probably would, as I did, have had an increased appetite, followed by certain improvement of health. Patients generally are improved by *mere change*. This occurs when the change is from a better to a worse atmosphere, and when the distance is short, showing how infinitely great the chances are in the finest climate known on all the earth. Whilst many physicians advise their patients to go south, to Louisiana, Florida, Cuba, eminent physicians of these localities have advised *their* patients to go beyond the tropics ; to spend their winters even far north in Europe, or the United States. All changes are good ; the greater the change the more probably it will change morbid action and cure all manner of chronic disease. The chances for permanent recovery are multiplied by the distance and number of changes.

A gentleman of St. Louis came to Colorado Springs last Fall. He appeared to be little else than a breathing shadow. Ten days thereafter I found him quite cheerful; said he had gained 3½ lbs. In a few weeks he declined. Although a gentleman of rare intelligence, I could not persuade him to make a succession of changes, or even one more to any other point, except homeward, he said, "to die." Now, with all due regard to candor, allow us to ask what would have been the accumulative result of a continuation of equally fortunate changes, assisted by the free air of elevated places, instead of a close room in a hemmed-in valley, near a mill creek, and surrounded by a profusion of putrefying animal and vegetable matter, as at the place he stopped? Moreover, here he had time to think of the dear loved ones at home, of whom he spoke so touchingly; whereas if he had been "on the wing," still improving, cheering prospects would have served more than as auxiliary to the cure.

Dr. TINGLEY, whom we prevailed on, about the same time, to go southward, is doing well.

The medicinal ingredients of celebrated waters are not, as generally believed, directly tonic, but indirectly so; they brace the system by improving digestion. All waters, except pure water, (that is, water distilled through the atmosphere or a vessel,) possess this virtue more or less according to the quantity of earths they contain, or the length of time they have been used.

We find that during well-directed journeys if the new excitants are too powerful the effect is easily controlled by substituting distilled water and the use of little or no medicine. Then they may be cautiously resumed. Also, if changes are not quite adequate to the necessary stimulation of the liver and other glands of digestion, fractional and alternative (not purgative and weakening) potions of mild medicine, mostly vegetable, as podophyllin, leptansin, iridin, &c., are the only auxiliaries necessary to that vigorous digestion which must be the foundation of lasting recovery from every chronic disease.

The effect of changes, in many respects, should be carefully noted in order to realize the fullest advantages of them.

Absence of knowledge of the manner in which, changes affect invalids, especially want of knowledge of the necessity of *frequent change*, until the cure is fully made, has caused a wide discrepancy of opinion on this great subject. Improvement, loss of improvement, and want of continued improvement have been ascribed to any and almost every other than the real cause; not unfrequently to the opposite of the influences which worked those results.

After examining all opinions previously entertained on this heretofore unsettled subject, I defy any physician to say *where* is the best place for consumptives, what class of patients should accept the low, humid atmosphere of the West Indies, where the disease is very prevalent among the natives; who need the dry atmosphere of high countries; who should seek an equable climate; and stranger still, who will be most improved by a variable, a changable climate. How the last condition conflicts with the yet generally received and unmistakably correct opinion, that an equable climate should invariably be recommended.

Had the disease been so protean in form as to require so many and diametrically reverse conditions for different individuals, a reliable climate would never have been found.

Since, digressively, we have infringed on another heading, " CLIMATE," we will add that the facts relative to proper climate cannot crowd the capacity of a nut shell. Here they are :

1. All changes of place or climate improve health, notwithstanding powerfully opposing influences, which may ultimately gain the victory, destroy life.

2. The purer the air, the drier and more equable the climate, the greater the chances for recovery.

3. The advantages of any climate, including all its influences, as air, water, food, scenery, &c., are lost by continued occupation, although they last until many permanent cures are made.

4. Great uniformity of temperature can be maintained only by migrating with the seasons in suitable climates.

5. Great altitude is of immense importance in the prevention and cure of consumption.

Many individuals lack the courage to start; a lethargic indifference steals over them which is fatal ; unaccustomed to journeyings, they look on it as an immense undertaking; but if they can only be induced to make the start they are surprised at the ease with which they get along.

We have dwelt mostly on the cure of chronic lung disease, trusting that the reader is aware that all chronic disease is an unit in a great measure. A cure, in any form of chronic disease, can only be effected through an improvement of the *general health*, and when this is done all local ills vanish. The same means we recommend for the cure of dyspepsia, asthma, bronchitis and consumption will be equally successful in other chronic diseases. Diseases of every organ, and even of the mind, yield only to an improved general health. Many females, doomed to fatal in-door-illness, could be restored to health and vigor by such a journey as is here proposed.

The effect of changes of food, and generous diet, so necessary to the restoration of health, must be closely watched. Whilst an abundant assimilation of nutritious food is necessary to a restoration from thinned blood and wasted flesh, and to give a clear skin, bright eye, roseate cheek, rotund form and an enlivened countenance, by the manifestations of renewed vitality and increased animation, all of which are conducive to a cheerful mind and a long and hopeful future, improper food, improperly prepared, may throw into the system elements of nutrition foreign to nature's wants, which will retard improving health, and may prevent entire recovery.

CLIMATE.

For the preservation of health, with a view to a long, prosperous and happy life, and especially for the cure of protracted indisposition or disease. get, first of all things, a pure and dry atmosphere. For want of information on this vital subject, vast numbers of the most intelligent people jeopardize their lives where they would not risk a penny. Could they have seen as plainly before them as they now can the past, some of the most fertile valleys of low altitudes, where there are more people of the present generation under than above the sod, would yet be unsettled, and many of the largest commercial cities would have had a higher and drier site, and the money used for funerals, cemeteries, etc., would have been used for drainage, ventilation, etc. They expose themselves to very great danger to realize property and drag out a miserable existence ; for at that age at which they should be in the most vigorous manhood, they have either passed "Time's curtained portal" or are wretched invalids, never realizing half the pleasures of life or half their natural days ! Whereas greater vigor, physical and mental, would have resulted from living in a healthy location, and they would have been enabled to realize many fold the amount, and twice the number of years to enjoy it, during which additional life their pecuniary effects, at ordinary interest compounded. would many times double.

To get a pure atmosphere, free from noxious effluvia, humidity, dust, &c., invalids must ever look beyond densely-peopled farming districts. Western Texas, once a very dry country, is becoming quite a rainy country. During the last few years extensive farms have been cultivated, and rains and dews become more abundant. Rains track the plowshare invariably. Before broken, a chance rain runs over the smooth ground, hurriedly, into the stream, and through it out of the country. After being plowed, the earth absorbs the water and slowly gives it up again in the form of vapor, causing a damp air, dews, and a *nidus*, around which other mists gather until a rain cloud is formed. Priests, in olden time, appear to have known this. When appealed to by the laity of thirsting nations for rain, they instructed them to disturb, as by beating, the waters, thus exposing for evaporation a greater area of water-surface, and getting a starting point for a rain cloud.

Although Texas, until the cultivation of the soil caused abundant rains, was considered a dry country, Texans ever observed that a west and north-west wind, coming from the Territories and Mexico, caused vegetation to wilt fearfully. So uniformly dry, even in the so-called rainy season, (from April to August,) are these climates that irrigation is almost entirely depended on for farming. In Mexico the winters are so dry, that no shelter is prepared for stock, and hay is loosely thrown on the tops of flat-roofed houses. The air is so pure that fresh meats, even the most perishable, as fish, &c., exposed to it for weeks. without salt, get sweeter. without any smell or taste of taint. which so quickly occurs in all humid and impure atmospheres, where humidity furnishes a vehicle in which come noxious gases and all material means of putrefaction and disease. Buffalo bacon is made, and kept sweet the year round, without salt. Independent of the purity of this atmosphere, it has, by virtue of its dryness, an antiseptic, antiputrescent, toning, bracing, constringent, or drying effect on the relaxed and issuing surface of the pulmonary organs, which is of *immense* moment ; and this toning power to the skin prevents that relaxation and weakness which cause exhausting night sweats.

This atmosphere is so dry that it shrinks wagons, furniture, woodenware, &c., which have stood the test of years in the American States; nothing but the most thoroughly dried timber for vehicles will answer here, and even then the tires, &c., need shrinking. Washerwomen find that instead of waiting for hours their linen is dried in a very short time ; and the penman, unless a very swift writer, has no use for blotting paper, the ink drying as fast as it flows.

It is difficult to express fully the pleasing and encouraging relief of mind and body afforded myself and others by a restoration from a whispering to a strong voice. This could not have occurred in an atmosphere the humidity of which constantly acts as a softening poultice to the lungs and the membrane which covers all the breathing surface. I was very agreeably surprised when I found I could "holler," and often amused myself when alone on the prairie by at first uttering very harsh sounds, and finally tones clear as those of yore— a great change of sensations from those of a few months before, when I was immoderately laughed at because of an abortive attempt to call attention to an exciting event.

A gentleman of strong and fully musical voice, after entertaining us by singing, said when he came to Colorado he could not speak above a whisper !

A superstitious stage driver asked me to hold his horses until, by moonlight, he could see what it was that would not move from the road. It was a dried wolf which some wag had staked up in the road.

Here appointments are not made on condition " If it don't rain." Never before did I have carriage harness in constant use all winter without any break, crack or rot, except at the bit-ends of the bridles, which were frequently wet by the animals drinking water. Carts and ox-gear, exclusively of wood and raw-hide, not a pennyweight of metal about them, when once dry give no further trouble. Not so, however, of so-called drouthy Texas, Indian Territory and Kansas. The belts of timber stretching across these regions testify to the precipitation of moisture during centuries. Beyond these, timberless wastes meet the eye of the agriculturist and greet the heart of the intelligent health seeker.

Water transparent as air, no timber to decay, no moisture to support first the life and then the putrefaction of myriads of animalculæ, of both air and soil, so common to humid districts; no dust from the farmers stir of the soil; not a gnat or a mote to be found in the air of selected camping places.

One-half the advantages of a change of climate cannot be realized by a sojourn in any one locality. This is more evidently true of northern than southern latitudes. The range of temperature within the latitudes of Colorado, with the changes of seasons, varies from 20° below zero to 100° above zero; whilst that of southern Mexico is far more equable—the summer being no hotter, and the winter exceedingly mild. Ice is seldom seen, and the so-called rainy season, which occurs in summer, having passed, the winter is dry and very enjoyable. Such things as damp night air, and risk of taking cold from sleeping with open doors and windows—or, which is more common there, out on the ground—are never heard of.

In northern latitudes, even in the most healthy climates, recovery from pulmonary diseases, during winter, is embarrassed by the necessity for artificially heated rooms and oppressively warm clothing and bedding, and the greater force with which morbid humors are thrown upon the lungs. By the use of such clothing and bedding, irritating matters, discharged through the skin, are retained in contact with it, until they are, to some extent at least, reabsorbed; and thus reaching the lungs and other diseased parts, aggravate the disease.

Examine a sunbeam, admitted into a sitting room, by a glass, or even by the naked eye, and see the millions of mechanical irritants brought constantly in contact with inflamed surfaces by respiration!

Fine and pure as is the atmosphere of Colorado, even during winter months, invalids cannot there avoid these irritating causes by out-door life in the winter and spring. The summer and fall months afford an atmosphere unsurpassed by any of earth—in the mountains and mountain parks unequalled—but the vicissitudes of winter and early spring are sudden and extreme. In any other atmosphere they would be unbearable. Men and beasts freeze to death every winter. One Tuesday noon, in January last, the thermometer indicated 63° Fahrenheit above zero; on the following Thursday morning the temperature had fallen to 23° below zero—86° variation in forty-two hours. Although in far more vigorous health than I had been for many years previously, I keenly felt the pressure of this and some dozen other severe storms, during the winter, on my lungs, and feared the ultimate result. But there I was compelled to remain. Railroads eastward and westward blocked with snow—I could not go in either direction; and even if I could have done so, I should only have obtained a milder, but certainly a damper and heavier atmosphere. I could think of Mexico—New and Old—but the season was too late and the distance too great to permit of anything else than a sojourn in Colorado for the winter.

2

Notwithstanding these untoward circumstances, the present population of Colorado, although largely made up of reconstructed invalids, wear more young faces and elastic forms under gray hairs than any people I ever knew.

During the very warm days, which alternate with storms, and destroy the equable temperature, the air is so dry that the snow does not melt, but is absorbed by the air. Drippings from the eaves, or very sloppy roads, are very seldom seen. These facts very greatly offset the objection to this fine climate on account of the great fluctuations of that season.

The most serious effect from these sudden and great winter changes of temperature result from those of night. Instead of growing gradually cooler—thus preventing that weakening perspiration which is the result of the relaxation caused by sleep—frequently, at this season of the year, during the night, zephyrs bring a warm air from the Pacific slope.

In this way, instead of gradually falling, as it does in this climate during summer, and in Mexico through the winter, the temperature of the air sometimes rises 10° to 20° towards morning, and one night as much as 38°; necessitating the hazardous risk of throwing off bedding, which, if retained, would cause relaxation, exhausting perspiration, severe colds and the want of long and refreshing sleep, so necessary to the infirm. Except when emerging from bed into the bracing morning air, a time at which all persons (unless it be those who have had night sweats, or are very thin in habit,) should actively take the cool air baths, any throwing off of bedding at night, or clothing during the day, is unsafe, unless very particular attention is paid to atmospheric changes. North of Mexico the advantages of camp life cannot be enjoyed during winter and spring. There, where a shower of rain may not be seen during the whole winter, that mode of living affords an exquisite pleasure, difficult to understand by those who associate inclement weather, with the idea of camping out, notwithstanding any information to the contrary.

As additional evidence that abundant vegetable and animalcular production and decay (for where the former is found the latter is certain to be,) and an atmosphere pent up by forests, are prolific sources of miasma and disease, we refer to facts known of the Cross Timbers, lower and upper, of the State of Texas. Most of these belts of forest are upon high, dry and sandy, rolling lands, where there is no stagnation of putrid waters; and yet the old settlers along these belts of timber know that the inhabitants within these timbers are very liable to fall intermittents, remittent and bilious attacks, whilst those out a mile or more on the prairies, if on high, dry and rolling lands, are not similarly attacked. Such diseases do not so frequently occur, nor in so severe a form, on high, rolling lands as in flat, marshy bottoms; and yet the same cause and effect are found everywhere in woodland districts, and in marshy prairies or open lands where vegetable production is profuse.

In high timbered lands there is stagnant air if not water, and the still air retains much humidity in it, and moisture in the ground. High winds on the prairies blow the moisture out of the ground and humidity and malaria out of the country. The plains of the Territories and Mexico admit of high winds and almost constant breezes; besides the soil is mostly a coarse granite sand, so that if rains fell, and were not blown out of the country, they would rapidly penetrate to many hundred feet through a sandy loom which has no clay foundation. I turned a volume of water nearly as large as my body, into a garden 50x190, all night. Instead of running over and wetting all of it, it sank down at one end—I sold out.

Nothing, since the discovery of vaccination, is of more immense moment to the invalid public, than the fact that foremost of all influences is humidity; next malaria, (I use the word malaria in a broader sense than that generally ascribed to it); *thirdly*, sedentary life; *fourthly*, the relaxation and depression caused by heated air; and *fifthly*, the stagnant atmosphere of low localities, as not only the chief sources of all manner of ills and aches to which flesh is *not* naturally heir, but they are constituted the herculean reapers of our race before reaching half our legitimate number of days.

Give us the muscle, physical and mental, sacrificed during any one generation of men by the destructive touch of these five influences—or even of the first alone, for without it the second cannot well exist—and we can people a continent, cope with all armies and revolutionize every government on the face of the globe!

Dry air, firstly, deposits no moisture on the person, and, secondly, rapidly absorbs that which is thrown on the skin by perspiration; and being largely a non-conductor of warmth, does not interfere with the required temperature of the body; whereas, in a damp air, evaporation from the skin is effected at an expense of vital warmth.

Fresh air is a luxury to the respiratory organs, and enlarged efforts at breathing are made instinctively when entering the open air.

The poor tenants of cellars and low, dark, ill-ventilated huts of narrow streets and marshy districts, die mostly of consumption, scrofula, white swelling, rheumatism, dropsy, or some form of chronic disease, and immunity from such influences bears a direct ratio to the different degrees of the wide range of gradation of predisposing causes which exist in every household, even the most favorably situated and best conducted. While such influences can force chronic disease upon persons not otherwise liable to it, those constitutionally predisposed may forever escape unless the predisposition is excited into activity by these morbific causes.

In all countries, whether very hot, very cold, or temperate, there is comparative exemption from consumption and chronic diseases generally, where the air is dry; and diametrically the reverse is true in direct proportion to atmospheric humidity. From the London Medical Gazette, vol. XXX, we get the following: "In Egypt the atmosphere is hot, extremely dry, and toler- "ably agreeable. In Australia the atmosphere is temperate and variable, but " very dry. At the Cape of Good Hope the temperature is high, very variable, " but very dry. In all these countries consumption is very rare. In the West " Indies the temperature is high, but little liable to variation, and the air is " moist. At Bermuda the temperature is moderate, and very variable ; the air " is very dense, but it is subject to considerable variations, and the amount of " aqueous vapors disseminated through it is great. In these countries phthisis " is very common."

Even in extreme northern latitudes, where the temperature is so low, during much of the year, as to precipitate from the atmosphere all aqueous vapors, there is astonishing exemption from pulmonary and other diseases, notwith- standing many untoward circumstances, especially that of suppressed action of the skin, and the consequent increased force with which excrementitious matter is thrown upon the internal organs. This, we insist, is easily under- stood, whilst all analogy, all evidence, and all reasoning on this interesting subject must satisfy, beyond the possibility of a doubt, every rational being that the same dryness of the atmosphere would be much better in a temperate climate.

The great safeguard to health and life, during midwinter, to the inhabi- tants of extreme northern latitudes, is the dryness of the air at that season in which the temperature is so low as to precipitate from the atmosphere all humidity. If the chilling power of a damp atmosphere were combined with that of the intense cold of these latitudes, it would be unbearable. The quantity of clothing, bedding, &c., necessary to comfort everywhere bears a closer relation to the dryness than to the temperature of the air. On the same latitude as New Orleans, in Mexico, and where the thermometer indicates the same temperature, the inhabitants are very comfortable in a single, very light, cotton garment, and tourists find a very light woolen shirt more comfortable than the inhabitant of New Orleans does his overcoat.

Statistics by Sir James Clark show a mortality from consumption in the West Indies of more than double that of London, which is sixteen per cent.; whilst that of Mexico is only seven-tenths of one per cent.; showing unmis- takably not only the comparative claims of each climate, and the fatal mistake of sending such invalids to the former countries, but demonstrating the soundness of our views as to humidity, malaria and relaxation.

The driest air, when pent up or stagnant about the body, as within doors, by heavy clothing, bedding, or to some extent, by interruption of atmospheric currents in timbered lands, becomes charged with humidity from the body, and trifling as it may at first appear, in its cumulative influence it is a prolific source of much mischief, and indicates the necessity of out-door life in open

21

space, and in a mild and equable climate, where breezes are among the freest of Heaven's gifts to man. Of two linens, equally wet, one in a room and the other outside, although there appears to be very little atmospheric circulation, the one outside will dry much the sooner.

It is strange that it should not sooner have been recognized that it is totally impossible to cure chronic disease where the atmosphere is constantly charged with causes of acute and inflammatory disease, which causes as certainly protract and continue chronic diseases as they produce fevers, etc.

Miasmatic causes of disease are now so well understood as to be acknowledged by all observers. Late one fall, after a rainy season, whilst approaching the Trinity river, from a prairie country so high that the timber skirting it could be seen a day's journey from it. I knew that when I reached that bottom, persons of pale faces and enlarged spleens, etc., would be seen. Just as premised, not a family had escaped. At many places there were not well persons sufficient to attend the sick. Three houses, near together, were shown, where five persons had recently died. This could have been otherwise by a location further from the vitiated atmosphere. Why so trifle with life? A squatter's sovereignty without health is as a blank on the face of creation. At the same time that prostrating disease had spread the whole length of this and other southern streams, on the higher and drier uplands were found the fair daughters and stalwart sons of frontier life—healthy, active, and of faultless form and feature, if not of toilet.

Through our whole course pleasing and diverting scenery, so important to invalids, presents itself in endless variety. The totally different aspect of the mountains from those of woodland states, the mountain scenes, different classes of animals, birds of different plumage and song, leafless evergreens, the beautiful mirage, mountains of ore of the precious and valuable metals, masonry of great antiquity, native fruits of delicious sweetness which only a dry air and bright light can give, geological specimens of great curiosity and beauty, and of inestimable value. These, conjoined with strange races of mankind, and numerous other attractions, play an important part in the cure of disease.

Before touching Mexican territory many evidences of progressive civilization will present themselves. Many flourishing towns, and farming, mining and stock-raising communities will be seen, where, as it were but yesterday, the silent solitude was first awakened from its primeval repose by the noisy bustle of the pale-face as he came to conquer this fair domain for the use of civilization. Until then the voices of nature's untutored children—the cataract and the storm, the scream of the eagle and the war-whoop of the Indian —were all that relieved, if they did not make hideous, both night and day.

On the upper Rio Grande, the Indian and Mexican inhabitants are gradually becoming Americanized, and proportionally more interested citizens. Where formerly both sexes and all ages concealed all the face except one eye, they first dropped the *rebosa*, so as to expose both eyes, and now the whole face, thus showing that they imbibe a spirit of emulation from the increased and improved population.

At Julesburgh, on the Union Pacific Railroad, we shall ascend the South Platte to Evans and Greeley colonies, thence to the Middle Park; reach the hot sulphur springs by August 10th, and here spend most of that month. Here excursions may be made to Grand Lake, the source of the Grand River, or to North Park, distance twenty-five miles. Here the North Platte takes its rise, and some rich gold washings are found. Through the Parks are many agate mines of great extent, fossils and petrifactions, whilst the mountains bordering on the Parks and cañons are rich in silver lodes, all affording pleasant and heathful pastime. Water-fowl, mountain sheep and a great variety of large and small animals everywhere at this season. By a circular hunt of ten miles radius, the Ute Indians (friendly) here killed, in August, 1868, over four thousand antelope.

All through these Mountain Parks are picturesque lakes, murmuring fountains and waterfalls, whilst through the cañons pass roaring mountain torrents, Elysian camping places along streams of ice-cold and transparent water; flowers and delicious fruits in profusion; spotted mountain trout—now declared the finest known—are taken by the cart load from the streams; two or three kinds of clover, luxuriant timothy, and a great variety of very nutritious grasses everywhere.

It is not saying too much to state that language is inadequate to a vivid description of this mountain scenery—no poet's pen or painter's pencil can picture it. In August, hundreds of acres together fairly groan under raspberries so juicy as to melt at the touch. Late in August we shall reach Mt. Lincoln. This is one of the loftiest peaks of the range, and was formerly known as Triqua, being the source of three rivers flowing into different oceans. The name was afterwards changed to that of the murdered President. Its summit, of easy ascent, has an altitude of over seventeen thousand feet. Here, within vision's ken, are scenes of wildest grandeur. Parks, rivers and valleys pleasingly beautified by belts of timber, alternating dells and natural meadows, all gemmed with streams and lakes of crystal waters, which ever reflect snow-capped mountains. Looking westward from the summit of this peak, the eye wearies in tracing seemingly endless chains of mountains, which stretch away westward towards California and southwesterly to Arizona.

At many places through these parks there are thermal, sulphur, chalybeate, soda and many other springs now growing into popularity, which may be carefully used for bathing or drinking; in many cases with great advantage. In September we shall reënter the plains through the Ute Pass, or natural ascent from the south to the parks. On this route it would appear that an

Omnipotent arm had been exerted to create a combination of great attractions. The petrified forest. celebrated geysers, or boiling soda springs. discovered by Fremont, and now being built up as the Saratoga of the West; enchanted or monument parks. gardens of the gods. in which rocks of less transverse base than height stand many hundred feet above the earth, appearing to vie with the world-renowned Pike's Peak, whose summit is distant twelve astronomical miles, make this vicinity extraordinarily romantic. In our next issue we will continue a description of the scenery of our journey. Our next camp, after leaving the soda springs, will be at Spanish Peaks, beyond the beautiful valley of the upper Arkansas; the second. on the divide of the Raton mountains; third. near Santa Fe, the oldest town except one in the United States; the next, quite probably, after a long journey, at El Paso Gap, through which the Rio Grande passes, and at which place we enter Mexico.

A pent-up and darkened atmosphere causes scorbutic affections as certainly as the air of cellars causes plants to be delicate and nearly colorless. Without change to *air* and *light* no known medicines could effect a cure in such cases. By availing ourselves of such a change we grasp the power given to us by the Omnipotent. All animal and vegetable life suffers, if. to any extent. deprived of these two freest gifts of heaven to man.

It is a matter very little understood. but we venture to assert is strictly true, that invalids should migrate with the seasons. through healthy climates in order to a cure of their ailments. A large majority can be cured by one trip. or perhaps by half of the round trip—leaving us at Jalapa. South Mexico, by midwinter, to return home via Vera Cruz and New Orleans or Havana to the east by steam; while others will be advised to remain with us and return with the wild goose of spring to the Rocky Mountains.

On the return journey our route will be changed. as much and as far as compatible with the object of the expedition. but even where the same line has to be retraced the spring scenery is so different from that of the fall that in many respects it will appear as though it were an entirely different country.

In this way an equable temperature may be secured during the whole year. and a mild and dry winter be passed in a pleasant winter climate. Light woolen clothing may be worn at all times. thus maintaining uniform warmth without excessive weight and without inducing exhausting perspiration.

The mild and agreeable temperature. so strongly recommended by physicians. can only be had by migrating with the change of seasons. It enables the skin regularly and fully to do its duty in freeing the system from offensive and irritating matter. Without such change, the skin, during winter, fails in a great measure to do its duty and the bowels. kidneys and lungs are proportionately overtaxed. and because of a greater sympathy between the lungs and skin most of the increased labor falls to their lot. This is especially true where disease, weakness or irritation exist in the lungs. Hence the impossibility of curing lung disease except under a combination of the most favorable circumstances. To effect such cure we offer the advantages of a winter sojourn far south of New Orleans, in a dry, pure and bracing atmosphere, totally unlike that of Louisiana, Florida or Cuba.

In this manner insensible perspiration is encouraged and the system gets rid of all effete matter. If sensible perspiration were to occur, then there would no longer be healthy action, but a leakage from the skin, tending to exhaustion, as do excessive purging, urination or expectoration.

The subject of free perspiration is very liable to catarrhal attacks: the clothing gets moistened, and evaporation from it and the skin is a chilling process. The chilly sensation stops all perspiration, sensible or insensible, and thus throws back upon the system the effete or worn out matter that would otherwise have escaped, to be again taken into the circulation and conveyed to the most irritable parts of the body. It is an impossibility to cure any form of chronic disease if the patient continues to take colds during the treatment.

One of the great causes of failure in the treatment of consumption in the American states is the humidity of the atmosphere, which deposits moisture on our clothing, bedding, house walls, furniture, &c. It is easily and entirely practicable, by the trip herein suggested, and by proper and judicious clothing, bedding, bathing, and exercise in the dry, pure and open air, to protect invalids fully against bad colds.

By constant exposure to a dry and pure air the susceptibility to take cold from it is greatly diminished, and by the adaptation of clothing to the vicissitudes of weather, *fresh cold*, so prejudicial in chronic disease, may be wholly prevented. Invalids frequently take cold without being aware of it, because instead of catarrhal symptoms being manifested as in healthy persons, it causes an increase of the symptoms of the disease under which they labor. That disease—whatever it may be—cannot be cured until the susceptibility to cold is gone.

During all the severe winter of 1870-71 (the hardest for many years in southern latitudes) we camped out, and although when we left San Antonio every member of my family had severe colds, from sleeping in close rooms, not one of us had the slightest symptom of cold after being relieved from the first attack. Animals in this climate, even, *though never cared for, yet never have colds.*

Our journey lies through a dry country, and at the time when we shall reach the various points on the route there will be generally neither mists, dew, fogs, snow, nor rain. The atmosphere is so transparent that small objects may be discovered and picked up at night. Among a thousand teamsters you will not see a torch, candle or lamp, although they hitch up and travel mostly by night.

As Captain Mayne Reid says in the "Scalp Hunters:" "I would that the subject of hectic flush could know this climate."

So dry, mild, and healthful is Mexico that travelers lie out in the open air at all seasons of the year; hotels are seldom found; vast numbers of the population lie on the ground as a constant habit; many counties together have not a physician; consumption is almost unknown.

Exercise, for the invigoration of the body and the cure of chronic disease generally, is prescribed by all authorities, but a suitable climate for regular outdoor recreation has not had sufficient attention. In the States of the Union rains are frequent, and sometimes continuous for weeks or months, so as to prevent *perfect* regularity in this most pleasing of all recreations, while other influences, growing out of this cause of irregularity, spoil the advantages of the exercise.

Exercise in mountainous countries gives fuller and more varied play to the muscles of the body than can be obtained in level countries.

A renovating process in the living body causes dissolution and replacement of tissue. The dissolved substance of the body is a deleterious matter in the system, and is cast out, chiefly through the skin, by means of perspiration. In order fully to rid the person of this substance it is necessary that the skin be cleared of it. Hence the necessity of migrating with the seasons so as not to have these humors retained in contact with the body by excess of clothing, or bedding, or pent-up air; and also to reach an atmosphere of sufficiently low and uniform temperature to give, without oppression by bedding, sound, refreshing sleep, so necessary to a recuperation of physical and mental vigor.

There is much want of analogy between a soldier's life and that of a well-regulated camp life, and yet from the former, many interesting deductions may be drawn. During the recent unfortunate war, the Southerners' were the worst provided for of any army of modern times, and yet, although greatly exposed to vicissitudes of weather, and suffering from great privations and irregular habits, mostly in unhealthy climates, after passing through contagious epidemics, they were proverbially healthy. Vast numbers of adults had not had measles, and measles and typhoid fever were epidemic in a very fatal form, during the first months of the war. Afterwards, remarkably good health prevailed throughout the army, and, notwithstanding unfavorable circumstances, many recoveries from chronic disease occurred; which recoveries could be traced *only* to change of habits and camp life in the open air.

Of all influences prejudicial to health, irregularity of habits is among the first to be conceded: and to this the soldiers were greatly exposed. Forced marches at irregular times, scanty, innutritious and badly prepared food, taken often in excess and irregularity; half ripe fruits, putrid water, impure air, scanty clothing and bedding, and great recklessness generally, were the common lot. Many slept with nothing between them and the frozen ground except one half of a thin blanket—the other half over them.

Then. as general good health prevailed in large armies, and many very remarkable recoveries from chronic disease occurred under such very unfavorable circumstances, the good effect being attributed to *changes* of habits and conditions, and to out-door life, it would appear that a well-regulated camping life, cautiously conducted, and in the most delightful atmosphere on the globe, must be the unmistakable means of cure for inveterate diseases.

Capt. Geo. S. D., now one of the principal business men of San Antonio, was entirely cured of chronic pulmonary disease by such exposure, if we may so call it. Not having time to correspond with him on the subject, and knowing his goodness of heart. and recollecting the free manner in which he frequently adverts to his very remarkable recovery, I take the liberty of using the initials of his name without his special consent. He said that his health was very bad—that he had expectoration of the most unfavorable character.—and that he was entirely cured by camp-life during the war, and has remained perfectly well ever since. He is now a very fleshy, healthy and live-looking gentleman—an embodiment of perfect health. Medicines did him no good, and had it not been for this accidental camp life, he would probably never have recovered.

During cold weather it is impossible, with in-door life, to keep all parts of the body alike warm, In common parlance, " one side, " to some extent at least, " is freezing whilst the other burns. " In well managed rooms, the temperature of which is regulated by careful notations of instruments, (which. by the way, is troublesome) this difficulty is greatly diminished, but to some degree it always occurs. In this way, colds, influenzas, &c., so frequently occur that they occasion but little surprise.

Nor are catarrhal attacks all. Pleurisy, pneumonia, rheumatism and consumption are the offspring of these apparently trivial influences.

Out of doors, air of an equal temperature touches every part of the body ; the changes are not so sudden as those of a room, in which the temperature may fall from 90° to 40° in the course of an hour. consequent upon a neglected fire, or by the entrance or exit of persons, instantly exposing those who are in the room to a freezing air. It is for these reasons that beasts do not have colds, consumptions. &c.

By some topical treatment as an auxiliary to an out-door life many diseases of women, in themselves as hopeless as consumption. may be cured. Those approaching. or in, the aptly called "critical period of life, " need not be ailing throughout the remainder of their existence. By improved and totally changed general health. they may safely pass that which would otherwise be directly fatal. or give a start to some lingering and ultimately fatal malady.

After the frequent, sudden and great storms of last winter, there was scarcely a family or person, in wide sections of the country, not affected with colds or some form of influenza, except Indians, Mexicans, and one company of white men, whom I found camped out, without a tent, and covered with snow, near Colorado City. If the whole population of Colorado had been in camp, such an epidemic of "bad colds" would probably not have been known.

Then it does appear that in-door life, even in the severest weather of the most rigorous climates, is as unnatural and pernicious as it is artificial.

When the inhabitants of the plains are comfortably housed, travel rapidly in close cars and stop at hotels instead of camping out, much of the good reputation for invalids of this fine climate of the mountainous regions of the territories even during winter, will be lost.

Although the catarrhal affections alluded to were so general, yet the attacks were much milder than those in the American States, very few causing fever, or terminating in pleurisy or pneumonia. In health, persons are apt to forget what they have suffered during sickness: therefore, any after-expression relative to such suffering conveys but a very imperfect idea of it. During this endemic we suffered severely, and having entirely escaped, the previous winter, by camp life, I almost determined never again to live indoors, and still entertain that inclination.

Whilst statistics show a mortality from consumption in many densely-populated parts of Europe and America of from ten to thirty per cent of the whole population, Mexico, so far as we can learn, has only seven-tenths of one per cent. of mortality from that disease, less than one to the hundred persons, instead of one out of every three to ten, as found in the first named countries! It is shown that about one-eighth of the whole population of England die of consumption, and the ratio of mortality from the same cause in thickly-settled and miasmatic parts of our country, of late years, is nothing less. These figures certainly are an argument in favor of Spanish America for such invalids, stronger than the same number of words in any other form could give.

Then if this is the country in which the fell destroyer, consumption, can scarcely raise his hideous head, it must unmistakably be the place of cure for those attacked. It has already so proved to be to thousands, and in the next few years will so prove to immense numbers.

Although the people of that country are indiscreet, and practice many bodily abuses; and although the poorer classes, constituting the great mass of the population, are raised on a very scanty diet, which in youth, when the plastic constitution can be moulded to circumstances, prevents the physical vigor and development which are essential conditions of great longevity, yet numerous instances of extreme old age can be found. In the same train may be seen the father, grandfather and greatgrandfather, members of four generations, all driving separate teams. A man over eighty years of age was pointed out to us, who briskly walked eight miles daily to the morning market.

Where is the physician or well-informed invalid, now-a-days, who believes that medicine is anything more than palliative in any form of chronic disease, and in a vast proportion not even that, whilst the patient remains in the same locality, breathing the same atmosphere, and subject to the same influences under which the disease was produced? Humane physicians advise such persons to travel, and if suitable arrangements, in a climate adapted to the purpose, had offered, such a course would no doubt have been the invariable suggestion of all physicians.

Recoveries do occur, it is true, but in such instances natural as well as artificial means play a part; some vital principle within the living economy co-operates with a favorable change of season or atmospheric condition, and the recovery is accidental. If the cure of chronic disease were as much under the control of the physician as inflammatory disease, the former would not be found in almost every family of our country.

From an editorial in the "Borderer," a paper published at Las Cruces, a town on our route in south New Mexico, we extract the following as evidence of the power of this climate in different forms of pulmonary disease: "If "any person could find an excuse for overpraising the climate of this portion "of our Territory, we believe we shall certainly be pardoned * * *. We "shall speak only of our own personal experience, and, if it shall be the "means of relieving the similar suffering of some of our fellow-men, we shall "be thankful indeed. For about thirty years we have suffered from that "terrible complaint, the spasmodic asthma; hundreds of times insensible, and "supposed to be in a dying condition. * * * In the fall of '68 we left "the beautiful State of Minnesota, to ascertain if a climate could be found "that would alleviate our sufferings, which had been so severe during the "summer that the only rest we had for weeks was by the use of chloroform. "Unable to travel we wintered in New Orleans without any benefit from the "climate. In the spring of '69 we crossed the Gulf, * * * arriving "at La Mesilla," (a village near Las Cruces.) * * "Here we spent the "summer and fall and were benefited, so that life became once more endur- "able. * * * We have learned to forget almost what the sufferings of "asthma are. We have climbed mountains all day on foot, taken long "journeys in all kinds of weather, without the slightest inconvenience. * * * "Of this fact we are certain, that it is the only climate in which we have ever "lived where the hearts of sympathising friends have not been saddened by "witnessing sufferings they had no power to relieve."

Warmth and moisture cause first abundant production and then rapid decay of all vegetable and animal matter. This warmth and moisture, and these substances of extinguished vitality, are abundantly found at all seasons of the year in Cuba, Florida, Louisiana, and other low and southern latitudes; and during the warmer seasons of the year throughout all of the woodland, rainy States and also in the flat prairie States; especially in densely-populated neighborhoods, towns, cities, and on or near streams and lakes; in fact almost any place considered inhabitable by an enterprising people. This putrefaction or decay generates, abundantly, malaria or miasma, which changes the whole atmosphere, and is inhaled in every breath, acting as a poison in the circulation, and causing many forms of disease. Chronic disease constantly keeps up irritation in the body, exciting to increased action the nerves, the heart and the arteries; thus wearing down the constitution, and causing that emaciation which is common to and characteristic of such diseases, especially of consumption. When inhaled, malaria gets into the blood circulation nature goes to work to cast it off. For this purpose the nerves and arterial system are taxed with additional labor, thus still further wearing down the subject of such excitement. Besides, the malaria lodges on the diseased parts, or somehow increases the disease it may have caused. Who would think of putting irritants into an inflamed eye? and yet, with what indifference invalids, by inhaling them, bring irritants in contact with diseased lungs.

Because of malaria, humidity, and the relaxation caused by the hot atmosphere of southern latitudes, or an impure, dusty and hot atmosphere, pent up by buildings, forests, etc., during the summer months, in more northern latitudes, mighty hosts of invalids are being lost who could be saved by proper regimen, and very little medicine, in the pure, mild and bracing atmosphere secured by change of place, with the change of seasons, on the plains.

It is an established fact in physiology that absorption, wasting or taking up of the substance of which the human body is composed is in an inverse ratio to the tone of the body; or the substance of the flesh is taken up by millions of absorbing glands, and cast away in direct proportion to relaxation or weakness of the body—a process fearful and fatal to consumptives and chronic invalids generally. It is through this relaxation that so many of such invalids die on the approach of warm weather. Then, in view of these admitted facts, invalids should beware of a humid and impure atmosphere, particularly in hot localities and seasons, as they would the shade of the deadly upas tree. Some have improved, it is true, by a visit to Cuba, mostly during the winter, but then there was a compensating influence—that is, a change which more than counteracted deleterious influences. Anything worth doing at all should be done well and quickly. Invalids who at home were totally disinclined to any exertion, declaring that they were unable to either walk or ride, have been surprised at the buoyancy with which they could get about in this bracing atmosphere, and expressed surprise at their strong inclination to do so.

Of dumb brutes the domesticated are the only animals much liable to disease, and of these the horse is the most liable, although in the wild state on these plains a blind or diseased horse is not to be found. Amongst thousands of these native animals belonging to trains, accustomed to camping out in this pure and open air, instead of being pent up in a putrid atmosphere, over or near heaps of putrefying excrements, I have never seen a diseased horse or mule unless it was from injury in working.

The skin, being supplementary to the lungs in promoting a normal condition of the blood of all animals, must be considered in all its relations to atmosphere. Many animalculæ breathe through the skin. The instinctive disposition of man and beast to rush to free air when oppressed by asthmatic or other obstructions to respiration, seems of more than single significance. An old and rational suggestion for restless nights, when oppressed by bedding, or, rather, by transpired humidity and humors retained in contact with the body by bedding, is to rise, walk around the room thrice and return to bed, when sleep will follow. Then, although relief is only temporary, and unsound rest follows from the cause which produced the first trouble, the experiment is satisfactory, and the unpleasant dreams in the after part of the night are not ascribed to the first and real cause. The skin not only eliminates from the body excrementitious matter, but assists the lungs in decarbonizing the blood. A circulating atmosphere is necessary to keep it washed of these humors. An impervious application to the skin will kill an animal in a few hours, death being preceded by great difficulty in respiration. Because of the humidity of the atmosphere of rainy countries and seasons, this perspiration, with its load of impurities, remains in contact with the skin, and is often absorbed into the system, and causes nervous disturbance and disease. A dry atmosphere rapidly absorbs this poisonous effluvia. Experiments very ingeniously, ably, carefully and successfully conducted, accurately to ascertain the effect of air of different degrees of humidity as effecting elimination of effete humors through the skin, have shown that during a given time this transpiration is six times greater in dry than in damp air. Because of this, more than all other properties of dry air, it is curative of chronic disease.

Many believe they do not sweat in this atmosphere, whilst those of an atmosphere charged with humidity think themselves subjects of profuse perspiration, the skin being often bathed in it, because the air, as a filled sponge, cannot take it up. Medicines are readily introduced into the system through the skin. These morbid humors also enter often and spread broadcast mischief.

Bathing, equally essential in the preservation and restoration of health, is a powerful means, and as such must be used with more discretion than generally believed. In health its injudicious use is well borne; therefore the impression, too general, that it is harmless. In sickness the susceptibility to its strong impression is usually very keen. No two persons require water of the same temperature, nor does the same patient, during the different stages of cure. Water, a little too cool, too long continued, or too often repeated, abstracts vital warmth-life. Much depends on age, temperament, habit, and especially the *effect* carefully noted. No one, even under the most favorable influences, can rapidly recover without judicious bathing.

The area of the surface of the air cells of the lungs of an adult, in contact with which inhaled air comes, is about 2,000 square inches, nearly equal to two square yards of surface, and the aggregate capacity forty cubic inches ; so that if an individual respire twenty times a minute 800 cubic inches would pass in and out every minute, or 48,000 cubic inches, equal to 333 cubic feet an hour would be rendered unfit for subsequent inhalation. A half dozen such persons would, in twenty-four hours, use near 48,000 cubic feet of atmosphere, and this, mixed with the air of the largest building on this Continent, would unfit it for perfectly healthy respiration. Think of this, you who would, by tight rooms, doors and windows, &c., shut good health in, and bad health, bad colds, &c., out ; who would wear bright eyes, rosente cheeks, elastic, active, strong bodies ; who would prize a sound mind in a sound body : who would avoid an atmosphere slow in its effects it may be, but fatal in its tendency.

DIGESTION.

The condition of the stomach and digestive organs bears an intimate relation to the cure of chronic disease, and by journeying as we propose the powers of digestion and assimilation are invariably improved. The stomach is the workshop of the whole body, and there is the closest sympathy between it and every tissue of the vital economy. A bruise of the hand or foot, or other remote part, will, in sensitive persons, frequently cause vomiting. The stomach and liver are inseparably connected, and disease of the one cannot long exist without involving the other. When the liver is affected the kidneys soon become diseased, causing dropsy, rheumatism, &c., diseases which yield only to a restoration of healthy and vigorous assimilation. In almost all chronic diseases the stomach is primarily or secondarily affected—generally the former.

As a result of indigestion, vast numbers of frightful skeletons are to be met at every turn, who live out only half their allotted days. Indigestion furnishes morbid humors or imperfectly assimilated chyme, which, passing through the circulation of the blood into every irritable organ, joint or tissue, causes and continues all conceivable forms of chronic disease, besides affecting the general health, and by bad general health every local manifestation of disease is enhanced in violence.

The effect of travel on the digestion, and through it upon the general health, is truly wonderful. A little patient of mine, who was so delicate as never to eat fat meat at home, had not been long on the road before she said, " Please give me the fattest piece of meat in the box."

An eminent author on the subject says that indigestion exists in forty-nine fiftieths of all cases of consumption. How he arrives at so nice a discrimination as to find that the fiftieth case, as implied, is not thus affected is difficult to divine. Indigestion exists in *every case*, not only as a symptom of but as an exciting and continuing cause of consumption. Through the pneumogastric or eighth pair of nerves, a direct sympathetic union exists between the stomach and the lungs. Whatever affects one of these vital organs will, by reflex action, reach the other, and consumption cannot exist without involving both. As well may the clock-maker attempt to drive the machinery without a mainspring, as for the physician to attempt to cure consumption or any form of chronic disease, without securing and maintaining a perfect digestion of nutritious food.

Dieting, as recommended in chronic disease, is a very unfortunate and often fatal expedient. Although it impoverishes, weakens, and kills, it is proper and even necessary where there is feeble assimilative power. Food unassimilated is foreign to nature's necessities, and cannot invigorate. When by journeying the appetite is keenly sharpened, an abundance of rich but easily digested food is powerfully and certainly curative.

A properly regulated medicinal diet does not mean of late years what it formerly did, a starvation diet! By it we mean *nutritious food so far as congenial to the digestive powers.* Now, here is the nice, the difficult point of discrimination between proper quantity and quality of food, and the mode of its preparation for different patients, and the same one at different times. No physician can invariably and correctly fathom this difficult problem. After cautioning the invalid to take only such articles and in such quantities as advised, and only at regular times ; to eat slowly, chew the food finely, take no fluid during or soon after each meal except such as is prescribed, never eat to fullness, to sit up after eating, guarding against immediate exertion, &c., the sensations caused by food in the stomach are to be consulted. In this way an observing physician or person may soon ascertain the proper quality and quantity of food, and being familiar with the chemical relations of different articles of diet, can from time to time vary it as change may require. He will discover that often a much deranged stomach can be restored to a normal condition, even by a full meal, of suitable and nourishing diet, just as certainly as it is known that the stomach and its contents may be converted into a fermenting vat by a morsel of unsuitable food or drink, or by mental depression.

Indigestion often is so troublesome, that persons nearly starve to death for want of knowledge and a supply of proper food, and this, too, whether they take much or little food : for if unsuitable the starvation is often greater than if they take nothing.

All wasting or chronic diseases, especially consumption, (which means wasting), must be treated by rebuilding the constitution by generous diet. The trouble is to get nutrition into the blood without overworking the stomach, probably already very much weakened. It is on this principle only that cod liver oil ever did any good. Easily assimilated without much effort of the stomach, it counteracted the emaciating course of disease.

The morbid appetite incidental to chronic disease, and especially when stimulated by favorable influences, is not an unerring criterion by which to be governed in the selection of food proper in quality, quantity, or mode of preparation. The wear on the system by disease creates a demand to replenish the wasted tissues. This demand often is so urgent, particularly with tourists, as to cause them to mistake the articles, and especially the quantity necessary. The food must be to some extent dictated by correct judgment, and *changed* as the different elements of support within the body are in excess or deficiency, which condition will be ascertained by an intelligent investigation of the appetite and of the effect of food on the digestive organs.

Camp life, in a very few days, gives a relish for food which no condiments can induce. Then and there is the place to correct any improper habits in the preparation and use of food. Digestion is so vigorous as to give full satisfaction without luring or urging by spices, cordials, &c., without which at home much nervous disturbance occurs.

The most horrible aspect to a fastidious mind is that of the angular outlines incidental to loss of rotundity of the human form by bad health. No artificial means can offset it, or reinstate juvenility; spent in journeying, less than half the outlay abortively used for this purpose, would prevent premature decay and the early loss of manhood and could not fail to give far more satisfaction.

The fact that abuses in diet, as from improper articles of food, or from excess in the use of that which was suitable, having caused weakness of the stomach, indigestion and distressing dyspepsia, with its long train of ills and aches, among a people who can afford an abundance of food, has erroneously schooled the public mind to an impression that to be cured, invalids must be dieted, or, to some extent, starved, when really the reverse is true, and in fact, without a generous nutrition, a cure is simply an impossibility. The whole secret of success consists in an increased assimilation of nourishing food, giving additional blood, flesh, strength, animation, warmth, *life!* For this *vital* purpose no means or combination of agencies has anywhere been shown or proven to be reliable, except that of out-door life, in bright and healthful climates, as that of the Italy of America. Local manifestations of disease are often, and in fact, generally, only signs of deeper seated and more extended mischief in the body, and can be relieved and cured only by the eradication of the wide-spread vice. Until this is done, consumption and dyspepsia, which. in our opinion, approximately, are synonymous terms, will continue, as heretofore, to be largely regarded as incurable maladies.

We find it necessary to dwell at considerable length on the relation which a vigorous digestion bears to the cure of chronic disease, in order to show *how* it is that traveling causes these happy results.

When the digestion is feeble, the "*fever of digestion*," *i. e.*, the warmth and life-producing process, in the living body, is low ; and in the absence of this required warmth, chilly sensations supervene. These, sometimes alternated by flushes of heat, or fever, are always followed by nervous disturbance, and an increase of the symptoms of every form of chronic disease. In order to continue, by a vigorous assimilation of food. this warmth, this life, this blood, flesh and strength producing process, until vigorous health is restored—a thing no longer considered problematical—it is necessary to keep up a lively sensibility to favorable impressions from new influences by frequent change. No change will be so quickly made, from improper habits in food, drink or medicine, as to disturb the nervous system.

After the restoration of a healthy digestion, and thereby the restoration of the general health. tubercles are readily absorbed before any softening occurs in the lung. In vast numbers of persons, from such fortuitous changes, tubercle has entirely and forever disappeared ; then physicians have been too much inclined to regard as arising from a mistaken diagnosis, previous opinions of the existence of lung deposits; and, as already shown, when the disease has advanced to destruction of substance, the resources of nature are immense collateral, and all auxiliary circumstances being favorable. This, too, has been in other and many diseases, still more plainly in accidents, considered hopeless. Soldiers and other patients of good constitution and general good health, have obstinately refused amputations of mangled limbs, at much risk to their lives, no doubt, but after much sloughing of flesh and bone, they have had sufficient use of such limbs to enable them to follow and guide the plow, and engage in other useful pursuits.

The one and only thing needful, as almost too frequently intimated, is an improvement of the health of the whole man, so as to invigorate and enable the vital, the living, the health and life-imparting principle within, to successfully combat all local manifestations of disease. Without this all is futile— all is lost.

From a very interesting annual, the Colorado *Gazetteer*, we get the following: "Whatever will aid the consumptive will aid the dyspeptic; for the consumptive is first a dyspeptic, and in fatal cases, *always starves to death*." By an unimportant little reversion of this sentence, causing it to say whatever will aid the dyspeptic will aid the consumptive, &c., our views are fully met. Also, from the same instructive volume, we extract the following: The writer. Dr. Wallihan, in speaking of the advantages of great altitude, says:

" The result is a permanent increase of the breathing capacity. The chest of a well-proportioned man, by actual measurement, has been known to expand three inches in as many weeks, after arriving here; and the appetite keeps pace with the respiration."

Many have told me, and showed me, too, that when indigestion caused spitting of thick mucous from the stomach, they expectorated similar matter, and that the quantity varied with that from the stomach.

So fully is it now known that pulmonary disease cannot occur above a certain altitude, that scientists have undertaken to establish a tuberculous zone, above which it is declared that the disease *never occurs.*

Hemorrhages of the lungs most frequently occur when the glands of digestion are torpid, in fact, very seldom, if ever, when digestion is good ; therefore it may be considered quite safe slowly to approach high countries, an increasing appetite keeping pace with such advance.

An eminent physician of California has recently said that observation and experience do not sustain the opinion, too generally entertained, of risk from lung hemorrhage in ascending great heights.

Consumptives having evening fevers, night sweats, or heavy expectoration, sometimes have a keen relish for food. This is caused by the wear of the fever upon the body, or by wastage, and is nature's mode of counteracting and compensating such wear and loss. If digestion under these circumstances ceases, or is much diminished, the patient will go down very fast. An opiate or other anodyne expectorant at this juncture, or an atmospheric or other influence which deranges secretion, has often destroyed, in a few weeks, the lives of those who otherwise could have been cured.

Persons unacquainted with chemical analysis of earth-waters generally, would be surprised at the quantity of chemical substances they contain. These, when new as a whole or new by reason of varied combinations, act as powerful stimulants on the glands of digestion, the same as does salt, pepper, mustard, wine, whiskey, vinegar, and all spices, condiments and appetizers when new. We say *when new*, because the power of all these substances is lost when the susceptibility to first impression is gone ; otherwise they would continue as at first, rapidly to give flesh until it would be an unbearable burden !

Indigestion, present in every instance of lung and most other chronic diseases, often causes the most wretched mental depression, disinclination to activity, avoidance of society, and to regard with distrust every proffered means of cure. During the hectic exacerbation, when the blood flows with increased force to the brain, in common with every part of the body, a species of intoxication is caused by the fever, giving a lively manner and temporary hope, but woful and direful is the corresponding depression which ensues. Here then is an object of commiseration, who should have the sympathetic encouragement of every humane friend, and be kindly urged and assisted to accept the only chance of life. Abundant will be the reward of those who thus may return to a distressed family a healthy, grateful and happy member of the household, again to fill the " vacant chair " in the accustomed corner. The possession of an empire could not afford greater satisfaction.

"The predisposing causes of tuberculous disease are all of a debilitating nature," indicating the necessity of a strengthening regimen from the commencement of the disease. Every influence which can invigorate the body, particularly by an increased assimilation of substantial food, must be looked to.

Poor assimilation leads to weakness, weakness to relaxation and prostration, these to more rapid absorption and consumption of the whole body, showing the necessity of the bracing mountain air, and the invigorating assimilation which it invariably gives.

MENTAL DIVERSION.

Long suffering and the gloomy thoughts engendered by reflecting on the possibility of an early grave, probably unknown to friends and relatives, induce an abnormal condition of the nervous system, causing not only mental depression, but the mind directs to diseased parts the morbid humors found in the circulation, which in their turn aggravate and increase the disease. By journeying the attention is diverted to something pleasant and important in prospect, a more healthful and hopeful train of reflections is induced, and thus the morbid humors are differently and naturally disposed of.

A volume of unanswerable evidence might be adduced, strong as the fact that mental depression has prevented the digestion of the most congenial food and caused black hair to become white, to prove the effect of the *condition* of the mind over *physical health*.

Moderate physical and pleasing mental exertion, stimulated by an exhilarating atmosphere and diverting scenery, give new hope, new energy, new strength and a more healthful action of the whole living economy. That effete or worn out matter and other irritating humors in the blood can, through the nervous system, by an act of the mind, be directed to diseased parts, increasing the disease and the pain, is known not only to pathologists, but to the people generally by familiar illustrations. The inflamed eye, blistered surface, rheumatic joint, &c., are made far more feverish and painful by friends frequently directing the patient's attention to the diseased part, as by frequent inquiries or remarks about it; whereas, if the patient's mind is directed to pleasing entertainments he suffers much less. Much of this may be caused by nervous exaltation as the result of attention being frequently called to the part, but this nervous exaltation invites with increased force the sanguinary circulation to the part, and with the increased blood comes humors for deposit, until an issuing sore, internally or externally, actually becomes a drain to the morbid humors of the whole body. It is immaterial which view of this subject we accept, or whether or not we blend the two theories; the cause and effect are the same, and show the necessity of mental diversion in the successful management of obstinate disease. Besides, somehow or other a train of depressing reflections wears down to emaciation the subject of them, thus causing that exhaustion which too soon characterizes consumption, and prejudices recovery from any disease. Hence the popular maxim " laugh and grow fat." It was to this Cæsar had reference when he said:

"Let me have men about me that are fat; sleek-headed men, and such as sleep o' nights: Yond' Cassius has a lean and hungry look; he thinks too much: such men are dangerous."

A fat Irish woman, keeper of the St. Charles baths, New Orleans, to an earnest proposition from me to give her a large sum of money if she would tell me how to grow half as fat as she was, replied, "you just go long and don't care for nothing"—on inquiry I learned that such were her habits. In an unhealthy atmosphere, through devastating epidemics, for probably forty years, she has ate, drank, slept and cared not for transpiring events, and therefore she is as broad as she is long.

Gray hairs are fit receipts to disease for its faithful ravages upon the mind and body. Many, after an improvement of health, the writer for one, have had their gray hairs changed to a mere silvery sprinkling.

The stomach is the centre of sympathy, and its power of digestion is weakened by unpleasant or strained excitement of the mind, or by disease long continued in any part of the body.

Leading authorities on such subjects enumerate as some of the causes of consumption, affections of the mind, as grief, disappointment, anxiety, late watching, or close application to study, without using proper exercise; confined and sedentary habits, especially in the impure air of cellars, factories, close rooms, &c.

Mental depression, bodily inactivity, and a want of new excitements to the mind and to the digestive organs often cause food to undergo a disorderly fermentation in the stomach, from which unsuitable elements of nutrition are thrown into the blood. These frequently act as fuel to the flame of disease, and then follows a series of drugging operations for that which is the *effect* and not the *cause* of disease! Abstinence is enjoined when the reverse is indicated, with the necessary new excitements vigorously to dispose of food.

Many inflammatory attacks have their origin in a shock to the nervous system, some sudden disappointment, or sad bereavement, or an over-strain by mental labor. Then a short excursion up the Hudson, or other point out of our great metropolis, has prevented a shorter but more protracted visit to Greenwood Cemetery! Slower, lower, and longer continued indisposition, less attended with perceptible fever, but which more fully and more unmistakably give the feature of a fit subject for a journey beyond Time's dark river is too often the spoiled fruit of continued broodings over misfortunes, over a hopelessly diseased condition, &c. Then it is that longer continued trips, a succession of excursions, alone can effect a cure.

38

EXERCISE.

Sedentary life in close and small apartments dwarfs the chest and the organs of it, and greatly diminishes breathing capacity. This unfavorable result has a direct relation to the amount of in-door life, and the dimensions of apartments. In quite circumscribed localities, owing to the topography of the immediate vicinity, the effect of different degrees of humidity, impurity, and stagnation of the atmosphere is well marked. Observers of those influences can go into the towns and cities and densely-peopled neighborhoods, without any knowledge of the whereabouts of the greatest number of consumptive, rheumatic, scorbutic and chronic diseases generally, and point out their abode ; whilst with equal celerity they can direct attention to the higher, drier, and lighter parts, where protracted disease is proportionately rare. Still the more fortunate denizens of the latter localities are only on the half-way ground to complete protection.

The vital necessity of exercise in the open air for the preservation of health and life, and the cure of protracted disease, is universally conceded. Life insurance companies now make it a condition of policies. It is dictated by nature ; the muscles were made for use, and without it dwindle away, and with them the whole man, physically and mentally. Unquestionable argument on this subject is so abundant as to make superfluous much from us, except with reference to new fields of recreation for the afflicted. Persons not able to bear even horseback exercise are improved by the gentle motion of a spring ambulance in the open, pure and bracing atmosphere in which we shall live ; and they are soon able to mount a horse, cautiously feeling their way in this most healthful of all moderate exercises by gradually extending the duration of their rides until they are surprised at the number of hours they can sit in the saddle, and be refreshed and strengthened thereby instead of wearied and weakened.

Much depends on the shape of the saddle ; saddles of broad seats, as generally made, put on a strain the muscles and tendons of the hips and thighs, which should be kept in a more direct line with the body. A cheap, half-rigged and narrow seated saddle is far the best ; such as are called, I believe, "The California saddle." For ladies, too, the smaller the saddle the better, provided it fits.

With much difficulty, I succeeded in getting a very feeble gentleman to take horseback exercise. He declared he was "not able to *walk*, much less so to *ride*." Finally he accepted, in somewhat a fretful way, an invitation of a cattle driver to ride with him, stating that he was "acting on the doctor's judgment rather than on his own," (which, by the way, was certainly right, *provided he had observed instructions*). Soon a bull chase presented, of some two hours' violent running over rocks, mountains, ravines, &c. Result—a restless night : next morning general soreness of the whole body, with protestations against again mounting a horse—" felt as though he had been through a bark mill." Some fever, increased cough, &c.; but having committed myself in so important an undertaking I was not to be thwarted. After he had been relieved of symptoms rashly produced, and lest the cattle drivers should again catch him, I accompanied him on short and moderate rides, gradually lengthening the time of that exercise. At the end of three weeks he rode twenty-eight miles in six and-a-half hours. I remonstrated by telling him I feared a recurrence of the first trouble, by an over dose of the best medicine known, too hastily taken. On calling, however, the next morning, I found him quite comfortable, eagerly awaiting an order to mount his horse. Such reports, apparently commonplace, are of vast moment to sedentary invalids.

Instances of the salutary influence of such exertion are familiar to all observers. Take for illustration the pale-faced female of in-door life on an hour's evening ride; note the quickened manner, increased flow of blood through all its crimson channels, giving a roseate hue to the complexion, animated countenance, sparkling eye, vivid wit, cheerful manner, merry ringing laugh—in short, an illumination of the whole being and stimulation of the secretions of digestion, giving a zest for an evening meal which otherwise would be stale, unpalatable and indigestible. How great must be the cumulative effect of such health-imparting influences, continuing through many months and thousands of miles, in countries where local causes of disease are scarcely found.

Inactivity is fatal to all persons, whether in vigorous or feeble health. Confine to his bed the healthiest person to be found and soon debility, disease and ultimately death will supervene: or, just take one limb of such person, as an arm; pinion it to his side, and soon it will lose flesh, warmth, strength and power of motion. Then let invalids and the sedentary take heed. If such a course will kill a well person, much sooner will it an invalid. Of this, hundreds of demonstrations might be given—few must suffice. When young in my profession, a ferryman on the Big Black River thought to oblige me by telling me how to cure dropsy. He said that by chronic chills, diseased liver, and inability to work, he was once reduced very low, physically and pecuniarily. Dropsy came. A family to support and no means, no credit, no strength to do it! He knew that just across the hill there was game. He had a rifle, but not strength to carry it. The parents of effort, however,—necessity and affection,—urged him to attempt it. By laborious, slow and staggering steps he reached and procured the object of his desire. Gradually the game shied off, and he had to cross many hills to reach it, until it was a whole day's effort to procure subsistence. By night he had " walked off all the dropsical swelling," came home " weak, lean and hungry enough to eat a horse and chase the rider." But by morning he again would be swelled up

" tight as a toad," and could scarcely drag his feet along. Soon, however, he found his relief was more permanent; in a few months he had perfectly recovered, and was a very healthy man at the time of this narration. Had he had the means of supporting his family, quite probably he would have lingered along to an early death.

Some years after the above report, a dropsical negro was submitted to my medical charge. I instituted the best routine treatment of authority, with varying results; sometimes better, sometimes worse, until, discouraged, I lost sight of the patient. When at the plantation one day, I was told that Ephraim was very bad. In calling at his cabin, I found him swelled up enormously, confined to his bed, and almost helpless. Knowing not what to do or say, I mounted my horse and rode off, avoiding any one who might claim an opinion of me. My mind, however, was ill at ease on the subject. Soon the above dropsical case recurred to my mind, and I determined to call at my earliest convenience. Walking, in this case, however, was not to be relied on, for when previously urged and driven to walk, he would be found asleep beneath a tree in some grassy glade. His attenuated blood no longer stimulated his brain, and, as is too common and fatal to invalids, he had little care for either life or death. I recollected, however, that some one was constantly in the saddle to do the errands of this and two other plantations on the same estate. Next day I called and had him mounted, after much demurring, for a half hour's ride, with instructions to increase the length of the ride fifteen minutes daily, until he spent all day in the saddle. By this and an active course of alteratives, tonics, bathing and generous diet, he fully recovered.

An active, healthy, fleshy and hard-working gentleman, now of Western Texas, told me that, near twenty years ago, he had heavy expectoration, hemorrhage, great emaciation and all the symptoms of consumption. He went to Dr. Green, of New York, and tried all reputable means of cure. Finding his means running low, he started for the southwest. Arriving near his present home, penniless, and with no means of support, save a shot gun brought for the purpose, he camped out in the open air, lived on only what he shot; and now, although a very prosperous hewer of wood, he neglects not his almost daily hunting excursion. He told me he could run to the top of a mountain without fatigue in breathing. He firmly believes, and correctly so, too, that if necessity had not driven him to exertion in a better climate, he would have died.

A man of Habersham, Ga., paralyzed in all of his left side, determined to reach the only other surviving member of his family, a son on Red river—distant many hundred miles. Being very poor he started on foot and crutch. At first he made only two or three miles daily, gradually increasing the distance, until on his arrival there fifteen miles did not fatigue him as much as two did at first, and by the exertion of the lower limb, in hopping along on it. increasing the nervous and arterial circulation in it. the required warmth. elasticity and flesh were in a measure restored to it. while the arm, from which less exertion was required during the journey. still lay cold, emaciated, and powerless at his side.

Many will insist that out of bed they cannot bear more than the fatigue of a rocking chair. And they never will, unless they acquire it by effort. They who can bear a rocking chair at home, can bear a spring wagon to the base of the Rocky Mountains. There an atmosphere of unalloyed purity. an invigorated digestion, pleasing scenery, &c.. soon will give them a new-born hope. Would that we could dwell at length on this subject, which has no measure in the value of gems or precious stones.

Consumption is much more frequent in the left lung than the right, the right side being, as a general thing, more exerted than the left. through writing and other labor. Unusual exercise causes soreness, &c., to healthy persons, and much more so to the afflicted. Excess in exercise, like all other excesses, wears out the most vigorous constitutions, and yet it is the lesser of the evils, and being such, even those accustomed to strong physical exertion through long life, live longer than those of sedentary habits. A well-conducted investigation of this subject would show that those accustomed to moderate, daily, and *perfectly regular* exercise (all other circumstances being equal), are much the longer lived. After disease has become positive, any more than very moderate exercise is not prudent; but this is almost as necessary as food, for without it very little food can be digested. The increased demand for food amongst hard laborers evinces the wear and waste of the body, as, were it not for this demand, the exhaustion could not be borne. By it, although the digestive powers are taxed with labor of itself wearing to the economy, the loss is well sustained.

"Dr. Guy found in the close workshops of a printing establishment that "the compositors, whose employment is sedentary, fell victims to phthisis in "the proportion of 74 per cent.. to 31 per cent. in the pressmen. who, though "breathing the same air. and in every respect subject to the same habits of "life, differ only in the active bodily exercise of the press; and among the "same class of operatives the deaths from the same cause did not exceed 25 "per cent. in those who took exercise in the open air. From the same author-"ity, it appears that in single females leading a sedentary life—as book and "envelope folders, bonnet cleaners. seamstresses, &c.—the cases of pulmonary "consumption. compared with all other diseases. were three times as numerous "as among those engaged in non-sedentary domestic occupations. as servants. "housekeepers. and shopwomen."

This shows as positive that the disease can be prevented as is now known that it can be cured.

ALTITUDE.

To prevent the development of, and to cure existing consumption, one of the first and most important of all indications is to expand the chest and lungs. It is positively known that persons of large chests are not so much liable to dyspepsia and consumption, and that consumptives of large chests live longer and have much greater prospects of recovery than those of the reverse form; therefore what nature has failed to give should be acquired by natural means apparently provided providentially for the purpose. To expand the chest and organs of it, various means, mechanical and otherwise, have been advised and suggested; some of them very ingenious and probably advantageously used; certainly that of exercising the muscles of the chest by full respiration, the inhaling tube of Professor Green, of New York, &c., none, however, equal to the great emergency; whilst this vital object may be invariably effected by visiting places of great altitude. It is generally known that the pressure of the atmosphere at the level of the sea is so great as to weigh fifteen pounds on every square inch of surface, so that a man of medium size constantly bears a pressure of 28,000 pounds. A heat of 212° is necessary to the boiling of water. At an elevation of 12,000 feet, being that much nearer the top of the atmosphere which invests the earth, it weighs only 5 pounds per square inch, reducing the pressure on a man over 15,000 pounds, and, at still greater heights, on water so much that the curious have drank it from a boiling pot.

A man's chest comprises nearly one-third of the whole exterior of his body, so that in ascending from a low altitude to that of 12,000 feet, he takes from that part of his person a pressure of many thousand pounds, allowing it and the organs within greatly to expand, by reason of the elasticity of the denser air within, and the necessity of a larger volume of the greatly rarified air of this high altitude for the purpose of respiration. This sets the lungs to work with increased effort to supply a vital necessity. The air cells and the whole of the lungs are proportionately and greatly enlarged.

To prevent and to cure consumption the most important object is to prevent the deposit of tuberculous matter in and about the air cells. This cannot occur when these cells are kept well distended by full respiration—such respiration as persons cannot refrain from, although they may cause pain when at great heights.

As consumptive diseases advances, soreness of the involved parts occurs, causing more or less pain or uneasy sensations on respiration, and very readily the patient acquires a habit, voluntarily or involuntarily, of refraining from full respirations, particularly against yawning. The effect is extremely unfortunate, hurrying to a fatal termination a disease which, by fuller respiration, could be postponed, and which could be cured by such efforts at greater altitudes. Very few consumptives are aware of any effort on their part to refrain from full respiration, and yet it occurs in every instance, causing contraction of the chest and lungs, and finally much deformity, independent of that caused by wasting of the lungs. The air cells, closed or contracted, become abnormal bodies in the lung, acting at least to some extent, as a foreign body and irritant in the vital organ, and hurrying to a fatal termination a disease too prone to rapidly run its fatal course.

Invalids of contracted lungs, in ascending high countries, generally experience pain in the chest and difficulty of respiration, but when the necessary adaptation of lung caliber to the changed circumstances is established, which occurs in a few weeks, these symptoms abate, and the patient breathes freely and easily, obstruction to a free circulation of the blood and air in the lungs is removed, and with it the great risk of hemorrhage, so generally fatal, and caused by such obstruction. When this is achieved a great improvement of the general health and of the lung disease may certainly soon be expected. The sooner the obstruction to the free circulation of blood in the lungs is removed by the gradual ascent to the rarified air of great altitudes the safer it is for the patient.

So long as the blood is obstructed in its free course by this contraction, there is risk of rupture of the capillary tissues, and consequent hemorrhage.

In the proposed journeyings, when ascending from the plains to great mountainous heights, pulmonary invalids will be slowly advanced, so as to gradually accustom them to the change, and the consequent expansion of the lungs. Until towards fall the line of travel will be near railways, so that if at any time the altitude should prove too great for any of the patients, they could, in a few hours, be very readily sent to lower countries, there to remain a few days, and then, by slow advance, rejoin the main body.

Our whole proposed course through the territories and Spanish America lies at a high altitude. Growing persons, of shallow or narrow chests, would do well to avail themselves of this rare opportunity. The chest and the vital organs which it contains can be so enlarged in young persons as to forever protect them against the development of a disease to which they may be constitutionally or hereditarily predisposed. There are now on this continent one hundred thousand persons, emerging from childhood to adult age, who will have consumption of the lungs and die of it, unless they avail themselves of advantages such as are here offered.

It will be our aim to make the party one of pleasure as well as of health. Mrs. Fullerton with her children and their governess will accompany us, and from the encouragement already received, there will no doubt be sufficient ladies in the company to form pleasant and agreeable society.

In approaching greater altitudes the necessity of more oxygen than is furnished by the rarified air is recognized by the vital organism ; and to meet this necessity, hurried respiration sets in ; pulsation keeps pace with the increased breathing. and this increased labor gives an increased appetite to compensate for the wear and waste of tissue.

In approaching greater altitudes, the necessity of more oxygen than that furnished by the rarified air is recognized by the vital principle, and, to meet this necessity, hurried respiration sets in ; pulsation keeps pace with the increased breathing, and this increased labor causes an increased waste and wear of the body, which is hazardous to feeble invalids, especially consumptives. Soon, however. this living principle recognizes the necessity of increased reparative powers, and with this comes the better appetite, increased assimilation and more effective blood and muscle forming processes.

It is neither desirable nor possible to reverse these ultimately fortunate results ; it is the wisest and, in fact, the only safe course to gradually accustom the invalid to these changes. This can be done without any risk on the line of the route, where, if necessary, a patient may be sent back in a few hours to a lower and denser atmosphere.

By a gradual approach to the mountains across the plains, it may be ascertained who can bear a greater altitude. The advantages of going into the cool and bracing atmosphere of the parks, as the heat of August approaches, will be of immense value to those who can do so without risk. Any who cannot there venture will be furnished with skillful medical aid at the towns along the foot-hills, or by some one of the many physicians who it is expected will accompany the expedition, until they can again join the train at Colorado City. On our return the lungs will have sufficiently healed for greater heights, without risk.

All persons who have given any attention to the subject know that tubercle, in almost every instance, occurs first in that part of the lungs of least motion from respiration, i. e., the apex of these organs. Certainly here is indicated the vital necessity of exercising. as by the forced respiration of great altitude, exercise, &c., all parts of the respiratory apparatus, with a view to prevention of these formidable deposits. An eminent anatomist has recently said that " from the direction of the bronchial tubes and the inspiratory force, the current of air is chiefly directed to the base of the lungs, and that it is only on the deepest inspiration, and when the base has become full of air, that the apices can be completely distended." We venture the declaration, *and no time or observation* will ever show the contrary, that, all influences being the same, the ratio of mortality from the disease ever has been, is now, and ever will be, in direct proportion to altitude. The effect of altitude in the prevention and cure of disease is mechanical, and, therefore, free from objections to medicine-taking.

From a very sensible writer we get the following :

" Might not more be done in practice toward the prevention of pulmonary " disease, as well as for the improvement of the general health, by expressly "exercising the organs of respiration?—that is, by practicing according to some "method those actions of the body through which the chest is alternately in "part filled and emptied of air? Though suggestions to this effect occur in "some of our best works on consumption, as well as in the writings of certain "continental physicians, they have hitherto had less than their due influence, "and the principle as such is little recognized or brought into general applica- "tion. In truth, common usage takes, for the most part, a directly opposite "course; and, under the notion or pretext of quiet, seeks to repress all "direct exercise of this important function in those who are presumed to "have any tendency to pulmonary disorders. Yet, on sound principle and "with reasonable care, it is certain that much may be done in this way to "maintain and invigorate health, even in constitutions thus disposed."

Piorry, a distinguished pathologist, declared that by full respirations tubercle was absorbed.

REGULARITY OF HABITS.

The necessity of regularity in the times and duration of exercise is shown by the soreness and disquietude produced by any unusual exertion; which soreness of the limbs and muscles and nervous disturbance are invariably felt. more or less, by all persons, and are often attended by some degree of fever the day succeeding such unaccustomed strain.

Even if the air of the malarious states were equal to that of the plains in purity, the weather would not admit of great regularity of exercise, without which the good effect is lost.

On our journey the drives will be so made as not to interfere with great regularity of habits. The necessity for particular times for exercise, eating, sleeping, bathing, &c., is largely shown, not only by the dictates of nature, but by observation and experience. Men, beasts and birds of countries where the day is many weeks long have fixed intervals for rest from the wakeful hours, and when the accustomed time arrives they are as much inclined to rest and to sleep as if night had come upon them. Many people have so accustomed themselves to particular times for retiring that they are asleep in a few seconds after abed, unless they have that day practised some irregularities, and the force of habit is so great that they scarcely can resist it at the regular time. A person sleeping at an irregular time of any day will be very dull at that hour of the next day, and very strongly inclined to sleep. The mother of a sickly infant, although a very healthy woman, will rapidly wear down by the broken rest incidental to caring for her child. Statistics of mortality show a larger proportion of deaths, according to age, among physicians than most other classes of men; although generally a prudent people, they are exposed to great irregularities of habit in the practice of their profession, particularly in sickly seasons.

REMEDIES.

Physicians generally recommend, as the first and most important of all indications in the treatment of consumption, an equable climate, and the advantages of such a climate are equally great in the cure of every form of protracted ill health. This can be closely approximated by a migratory life, and only by so living. All opiates and anodyne cough nostrums taken into the stomach torpify the glands, derange digestion, and, although they give temporary and pleasing relief, are generally followed by injury to the health, which not unfrequently hurries disease to a fatal termination. These narcotics relieve and *appear to do good*, only by deadening sensibility. They blunt sensibility, for the time render the patient unconscious of the irritation which, under ordinary circumstances, gives rise to cough, and not only derange the general health but often prevent necessary rejection of irritating humors by coughing. When the cough is caused merely by irritation of the respiratory organs it is far better to allay the irritation by inhalants, &c.

Accumulated humors, phlegm, mucous, or matter, irritate the parts with which they come in contact, and cause a cough for the removal of such matter. Opiates and narcotic expectorants diminish cough and expectoration, intoxicate, please and will destroy that sensibility of the glandular system which is necessary to healthy secretion, and thus increase that indigestion which hurries the disease to a fatal termination.

Narcotics for the relief of irritation are salutary, were it not for torpifying effects on the glands; and to avoid this effect, it is far better to relieve by inhalation. Few, except experienced physicians, can tell when the cough is the result merely of irritation. There are few remedies regarded by the people as more harmless than cough medicines and cough nostrums, and yet there are not many capable of doing more mischief. We know that such remedies, injudiciously used, have caused consumption.

Medicines are a kind of necessary evil rather than the lesser of two evils. Judiciously used against acute disease, they hurry feverish action to a favorable termination; whilst a general want of success in the treatment of chronic diseases leads to empiricism. It is not possible to discover reliable means of cure for chronic diseases, whilst the cause of it, in the form of the patient's habits and surroundings, continues to operate. Such a proposition is an absolute absurdity.

Consumptives cannot recover at home. They must be emancipated from the influence by which the disease was produced, or where a constitutional or hereditary predisposition to it was developed, and especially from cough nostrums, the bases of which are narcotics—chiefly opiates—which intoxicate, relieve and please, but slowly kill by paralyzing the secretions; thus deranging digestion, and vitiating the blood, and through it the entire organism.

47

The fact which is now unquestioned and unquestionable, that stimulation of these secretions, by change of water and other suitable excitants, cures the diseases, is positive evidence that it is certainly fatal to torpify them by narcotics, which temporarily relieve the cough and pain, but ultimately lay destructive hands on the fairest and noblest structure of divine creation.

Turn to any work on the medical properties of substances, for instance, the U. S. Dispensatory—which is to be found on every druggist's counter—and you will there find that "Opiates *diminish* or *suspend* all secretions, with the exception of that by the skin," thus preventing the issue of offensive matter by the bowels, kidneys and lungs; a property of inestimable value when judiciously used against excessive depletion, but often fatal when used by patent medicine venders for the pleasing and temporary relief produced. Any mode of treatment, without a total change of climate, will be futile and often baneful. Croton oil, tartarized antimony, and other articles, used on the side or chest, as irritants, have been absorbed and through the circulation reached the bowels, thus bringing on diarrhœa, so common, so prostrating, and so fatal in consumptive cases.

In many bronchial and lung diseases, patients should have, in conjunction with and as auxiliary to, a camp life, inhalations, sometimes of a soothing, and in other cases of a healing nature. This is a mode of treatment unlike almost every other heretofore proposed and practiced, which will gain instead of lose favor the more it is known and judiciously practiced. The breathing surface of the lungs being nearly equal to the surface of the whole exterior of the body, is sufficient evidence that if inhalations have any virtues they must be of vast importance. Many years ago it was accidentally seen that persons with lung disease, working in the sugar mills of Louisiana, improved very much. This is easily understood when we recollect that sugar—chemically speaking—is very largely a carbonaceous substance, and, like carbon or charcoal, a great antiseptic, antipurescent and absorbing substance. Sugar, because of peculiar combinations, has a more restorative power over putrifying meats than carbon, salt and other substances used. Sugar applied to foul ulcers on the exterior of the body, changes them very fast to sweet, clean, odorless sores; and, if the general health is good, to granulating and healing ulcers. Had it not been for the warm, humid, relaxing and malarious atmosphere of the sugar houses of Louisiana, Dr. Cartwright would have obtained a world-wide reputation for the cure of these diseases by this vapor. Sugar, in suitable form, and its kindred substances, carbonaceous in their constitution, are the inhalants where there are issuing sores in the air passages.

Another and more important class of inhalants are anodynes to allay all irritation in the air passages, for as long as irritation continues in these parts no stage or form of consumption can be cured. During the late war, I, through a humane motive, accepted a contract from the Surgeon-General of the Confederacy to grow indigenous remedies. Opium, because of the block-ade, had advanced from six or eight dollars to many hundred dollars per pound, and there was much suffering among the people for want of it. Having been much exposed to heavy and cold dews, I was attacked by a very severe catarrh, which extended to the lungs. But my crop of these plants was in a state of ripeness for attention, and with fever, pain and soreness in the chest, cough, &c., I spent most of my time bending over these plants, and working with the drying and evaporating juices of them. At the end of the first day I was so much relieved as to be induced to place a portion of the gum near my face that night. Next morning I awoke nearly well, and soon was entirely cured, although continually exposed to the inciting cause of the disease. I then either took to my medical garden or subjected to the vapor of these drying juices every available person in any way troubled with disease of the respiratory organs, and with the most gratifying results. One, a negro, then belonging to Mr. Curtis, a merchant of Spring Hill, Alabama, who had all the symptoms of tubercle in both lungs, entirely recovered. I found that it was only while drying that these substances gave the curative vapors in sufficient quantity; and I have ever since managed to keep these juices in a suitable state to use. Whilst drying, an ounce of these juices gives more medical vapor than a pound after dried. Analogy led me to a trial of many kindred anodynes, but I found nothing equal to the juice of the poppy and lettuce, the former being the most effectual and the vapor of gum-camphor being the nearest to these two. On the journey we shall be prepared with these substances, and with suitable instruments for administering them.

In view of the fact that miasmata and other morbific agents, diffused through the atmosphere, some of them so potent as to depopulate cities and countries, such as cholera, small pox, typhoid, yellow, scarlet fevers, &c., are so readily introduced into the system by inhalation, it is strange, first, that such influences have not been more avoided in the treatment of disease; and, second, that suitable remedies have not been oftener addressed to the pulmonary surface. Often a lung or bronchial irritation, produced by an irritating or temporary cause, produces hard and harassing paroxysms of spasmodic coughing, which, if not checked by an anodyne inhalant, will keep up irritation and counteract the most salutary influences. Anodynes, properly used, by inhalation, do not torpify the glands and derange the health as when administered through the stomach.

Nauseating expectorants, so often used, causing relaxation and prostration, and increased absorption of the tissues, or, in other words, consumption of the whole body, are even more objectionable than narcotics. Often nauseants and anodynes are combined. Then who will undertake to predict the result, or to express surprise that lung diseases are so generally fatal? In this climate, under judicious management, the expectoration, (which, when prevalent, is equal to a loss of many times that quantity of blood) gradually diminishes, and blood, and flesh, and strength, and a cheerful spirit are proportionally restored.

Irritation of the respiratory organs invites, with increased force, the circulation, and in that way no doubt brings to the lungs the material for tubercle. Such irritation and cough ought to be checked, but remedies should be used with a full view to the fact that if the cause of the cough is morbid, corroding humors, which ought to be moved, it is dangerous to check such effort by nature. The difference between these causes of cough can be readily seen by the necessary attention to the subject, and by way of encouragement to those thus affected we must say that remedies locally used by inhalation will not interfere with expectoration, as do those which intoxicate the whole body.

Independent of and aside from all other evidence on this subject, the invalid will readily feel that as soon as digestion is invigorated he improves, and is far more cheerful and hopeful.

The total impossibility to cure chronic disease by remedies of the regular school of medicine, as long as the cause continues, has led to a trial of every proffered means of cure ; and if the patient, perchance, happens to improve by virtue of some favorable change, external or internal, then dame nature gets none of the credit, but that identical remedy, which at the time was being taken, although it be in portions too small to affect the most minute insect.

How the impression ever got so general among a very intelligent people that *manual* labor or machine power in the hands of man could have an effect which the perturbations of nature could not have without destroying all terrestrial life, is as strange as it is true. It is declared and accepted that by agitation or shaking, by pulverizing or making fine, by trituration or bruising, &c., the power of medicine is increased a million fold ! Well, if a million fold, certainly by additional shaking, &c., it can be increased a billion fold, and if a billion, infinitely. I say manual power because if the agitation caused by all the commotion of nature had a similar power, medicinal substances afloat on the earth would be given such potency that annihilation of all vitality would be an unavoidable doom ! Creation would be a blank ! When storms disturb the earth, upset strong foundations, shake in twain the mighty monarchs of the forest, and earthquakes shake down mountains of the eternal rocks, let such believers stand appalled !

An affable, plausible gentleman, of that school of thinking, once came along through a community of prosperous people, and in a very showy way ; soon he had patients in almost every well-to-do family, especially amongst the ladies and children, although the country, at the time, was remarkably healthy. Among his patients was a very large and healthy young lady, weighing over 200 lbs., never known in the neighborhood or by the family physician to be an invalid. He gave her the size of a mustard seed of ——! How ridiculous it looked for such an immense person to take a quantity of something which, if removed from the gum which contained it, could not be seen by a microscope giving five hundred diameters. On their theory that anything which will produce a disease will cure it, she must have taken the millionth part of a grain of pepper or mustard ; or the billionth part of a drop of whiskey or of Saratoga water to satisfy that which these articles give—an appetite, as well as and instead of food ! Great economy, certainly, in household outlay.

Suppose a person has taken, by mistake, or otherwise, a large quantity of morphine which will certainly kill. A regular physician present declaring that black, strong coffee—coffee in very large quantities used internally by the mouth, by enema, and by immersing the patient in it, will as certainly destroy the poison as water will extinguish fire ; and that the patient must be kept awake, even by flagellation if necessary ; whilst a homœopath would say, "Tut! tut! away with such inhuman treatment; here, take the billionth part of a drop of laudanum, or some other substance which would cause similar poisoning, and you are well." Which remedy would be chosen? If the family were homœopathic, I suppose they would choose the treatment of their physician ; if so, away would go the patient !

Through no ill-will whatever have I thus written of this class of so-called doctors. The spirit in which I have said so much on this subject is, I trust, far more laudable, and humane than that of envy. No one of this profession has ever given me offence ; on the contrary, many have met me very kindly, whom I must regard as candid and conscientious, although I know them to be astoundingly in error.

Most expectorants contain some narcotic or sedative principle, and in proportion to such ingredient destroy sensibility, in that way diminishing cough and expectoration ; this, conjoined with a species of intoxication, has a pleasing effect and gives much satisfaction. But the matter retained burrows into a vital organ, and, increasing that ulceration diminishes the capacity for respiration, hurries the disease to a fatal termination.

CURABILITY OF CONSUMPTION.

The length of our days is, to some extent, in our own hands, and we are recreant to a Divine trust unless we improve every opportunity of saving our health and lives.

The least indiscretion or neglect of personal duty may sacrifice what would otherwise be a long, useful and happy life. Of this, abundant and unquestionable proof could be adduced. To only one instance of such sacrifice will we refer. A lady told me that hereditarily her family were consumptive, and I saw that constitutionally they were so. By prudent habits they prevented a development of the predisposition, until the late war. A sister, then in excellent health, received information that her husband would arrive on a certain day. Expecting him on a Hudson river packet, and there being no conveyance from her residence in New York to the pier, she hurried there on foot. Not meeting him she hurried back, hoping he would come by a train from Albany, then due. On reaching home, in profuse perspiration, she would not enter the house, but sat on the stone stoop—a good conductor of warmth. She took pneumonia, which developed into consumption, and died in a few months. Comment is unnecessary.

Acknowledgment of the curability of consumption is of modern date. Formerly it was considered incurable. Too many cicatrices from obliterated lung-cavities have been found on *post-mortems* of persons who died of other diseases; and too many persons now living and well, who previously had unmistakable symptoms of actual consumption, are met with, to any longer admit of doubt. And if the management of these diseases now takes the right direction, thousands of living monuments of such success will soon testify to the same happy results.

Development of the disease can certainly be prevented—in incipient form it can be cured—and in the more advanced stages, if not perfectly cured, it can be much relieved, and fatal termination greatly postponed.

As it is difficult, in many instances, except by an expert in the examination of such diseases, to ascertain the extent of the ravages committed by the disease, it is certainly judicious for persons to avail themselves of this chance of recovery, and if only improved, so that life is again enjoyable, it will be self-evident, that under more unfavorable circumstances, such improvement could not have been realized.

The day is close at hand when the facts herein set forth, and the theories herein advanced, will be established beyond possibility of doubt ; we hope, even now, before it is too late for those thus afflicted—and then these plains and mountains, and Mexico, will be filled by invalids from home and abroad, regularly migrating with the seasons, until, fully protected against, thoroughly cured or relieved of the disease, life will be sweetened and prolonged.

Consumptives are dyspeptics, and the blending of these diseases causes misgivings and doubts on all subjects—even of the existence of the disease. This is the greatest cause of the great mortality of the disease. It is very unfortunate for them that they do not recognize the existence of the disease at a time when it is as certainly remediable as it is incurable at a later period. In the incipient stage, great offense would be given if told, even by their physician, that anything serious was the matter. If a physician so declared, in this stage of the disease, and the patient recovered—which certainly would be the result of the course here advised—he would lose the most lucrative practice among the most enlightened people. And this unfortunate condition of public sentiment will continue until the people, generally, know that consumption in its fatal form does not come on suddenly, but is preceded by some little hacking cough or stitch in the side, which symptoms, it is too generally believed, ought to be relieved by some good mother's remedy. They will spurn, as much as they would an accusation of dishonor, any intimation of serious disease. It is proverbial of consumptives, even in the last stage, that they will not believe it serious; and this is far, very far more true of those who are curable.

Three years ago I urged a gentleman, then in the shoe house, 101 Canal street, New Orleans, to accompany me. He replied that his business engagements were such that he could not go. Poor fellow ! He quite probably felt family relations so morally binding as to place duty before life ; appeared to ignore the fact that immense possessions without health and length of days to use and enjoy them, were as mountains of precious ores without a prospector. Although at the time not near so far advanced in the disease as I, he gradually sank and in a few months died. I know as well as anything can be known, before positively seen, that he could have lived. Then attention to business would have been rational and successful.

The term consumption, as generally understood, too exclusively has reference to the advanced disease, when it, like other incurable diseases, had passed the curable stage, and heretofore has generally baffled treatment. A much greater number of consumptives are seen afoot, seeking pleasure and business, suspecting nothing more than a "slight cough," which they expect to relieve by some "simple (?) remedy," than are found of features of disease which meet the general acceptation of the term consumption. If these could see the seriousness of early symptoms, and accept of the now established means of certain cure, then indeed would this immense destroyer be regarded as strictly curable.

A necessity conceded by all physicians is that of removing the producing cause of disease in order to its removal. Then if we show that the disease scarcely occurs in certain climates, such are the reliable means of cure, so long as there is any chance for recovery.

The time at which there is the least fear, there is the greatest risk—not so much of an immediately fatal event (except by sudden hemorrhage), but of passing to a stage of the disease in which the most favorable means are, at least, more uncertain.

Consumption of the lungs is a complex disorder. The local lesions in the pulmonary organs are the production of a vice widely diffused through the body, which can be eradicated only by an improvement of the whole economy.

Dr. Parrish, an eminent physician of Philadelphia, at an early period of life was known by the most distinguished physicians of that city to have lung consumption. He entirely recovered, enjoyed vigorous health, and in old age died of another disease. In each lung were found cicatrices, giving absolute proof not only of the existence, but of the cure of the disease. Then this is strictly a curable malady notwithstanding the immense mortality from it, which, we must insist, is as much the result of a want of constitutional instead of local treatment, suitable hygienic management, and necessary climatic changes as that of virulence of the disease. The absence of any one of these advantages gives a very slim chance of recovery : of all, scarcely a hope ! The first crop of tubercles seldom destroys life. By correction of that habit which leads to such production, these are absorbed, and then by continued good health the cumulative and fatal tuberculous formations are certainly and entirely prevented.

Laennec, after examining the lungs of thousands who died of consumption, declared that he had seen many cases " in some of which large portions of the lung had been destroyed by consumption, but the part, nevertheless, had entirely healed."

Light, so essential to all life, (animal and vegetable) is partly lost during winter in northern latitudes, in forest countries, and everywhere by indoor life. Persons of scorbutic habits are mostly found in darkened habitations, and among the poor of large cities, often in cellars, and their complexion shows the necessity of the action of light.

The journey scarcely can fail to cure chronic bronchitis which, if neglected, is so generally a precursor of consumption that it may be considered an incipient form of that disease.

Asthma, like bronchitis, chronic pleurisy, scrofula, catarrh and clergyman's sore throat or chronic tonsilitis generally terminate in serious lung disease if not cured. Old ulcers, skin diseases, fistula, ulcerated piles, chronic dysentery, and issuing disease of the uterine and urinary organs may throw upon the blood circulation, by absorption, humors, which, reaching the lungs, cause fatal disease there. We take the risk of after censure by cordially inviting all such invalids to accompany us.

From a recent prize essay we get the following :

"The curability of consumption can no longer be reasonably doubted ; the "fact having been verified in numerous instances of persons who, after having "presented all the general and local indications of its existence in various "stages of structural lesion, have recovered—living in the enjoyment of "tolerable health until a more or less advanced age, when, on their succumb- "ing under some other disease, the examination, *post-mortem*, has revealed the "traces of the former malady. Laennec remarked this occurrence on several "occasions. MM. Ferrus and Cruveilhier state from the result of their obser- "vations upon the bodies of old men and women, who died at the large hos- "pitals, Salpétrière and Bicètre, that it is not uncommon to find excavations "and other consequences of tuberculous disease which had existed at a former "period. More recently, M. Beau stated that 157 out of 160 women who "died in his wards in the Salpêtrière had cicatrices in the summit of one or "other of the lungs, which he considered to be the remains of tubercular "disease; most frequently the summits of both lungs were affected. M. Prus "likewise found, on examining the bodies of old people, in a large proportion "of them, traces of former tuberculization of the lungs."

MORE ABOUT CLIMATE.

From observation we derive information of vital value. Every inhabited place is one of putrifying matter, and the whole earth round about issues a noxious effluvia, which, inhaled at every breath, has an immense cumulative effect. The breathing surface, of the lungs and the absorbing power of that surface are about equal to that of the whole exterior of the body. Why so willingly retake into the system offensive, yea, very repugnant matters both by respiration and by absorption through the skin, which nature so vigilantly, so abhorently casts off? An invisible gaseous form makes them none the less nauseous and noxious, but the easier of introduction into the system. As well retake the excreted substances of solid and fluid form as those of gaseous and invisible shape! The solid and fluid issues are the effete materials of the body, and so are the gaseous exhalations of the lungs and skin; and yet in close rooms the most fastidious people retake substances cast off as no longer fit for the support of life! An exhaled air, deoxygenized by respiration, is no longer fit to be retaken, even by the same individual, and therefore the Providential provision of a moment of rest immediately after each respiration, so that the expired air, being rarified and rendered lighter by admixture with gases, and the warmth of the body, may rise up and pass away, and the purer and heavier air rush in before an inspiration commences.

Eberlee—I believe it was—an eminent teacher of, and writer on, medicine, confined rabbits in dark, damp pits. In a few weeks he found tubercles in their lungs. In a few weeks longer a part of the same lot of rabbits, having been removed to a high, dry and pure air, were free from such lung deposits.

An impure, pent, and hot atmosphere invariably cuts short the natural period of life; and, therefore, is more certainly fatal than the virus of the rattlesnake, but much less dreaded, because in an insidious form—"a snake in the grass!"

Many remarkable cures effected by journeying have been ascribed to the reverse of the real cause—one instance: A distinguished physician of Philadelphia, knowing that dyspepsia of a "fast liver" of that city could not be cured as long as he remained at his luxurious home, thinking to starve him into a cure, assured him that he should entirely recover if he would pledge himself fully to follow instructions. Having fully committed him on his word and honor, he started the patient afoot into the interior of a poor, mountainous and thinly-settled country, with instructions to buy and himself drive sheep to Philadelphia. Result—*vigorous health.* This cure has often been referred to as the result of dieting in a country where food was scarce, when the opposite is positively so. True, he found not the so-called luxuries of his home, but those of a more substantial nature, which the new excitements, mental and physical, enabled him to take with a greater zest, and which furnished more nutrition. Instead of being starved into a cure he was fattened into such, as increased flesh, strength and animation showed. There is no disease of cities so fatal as dentition. Let the mother of painful vigils press to her aching heart the frail form of her intense solicitude, and fly to a climate where the disease is not known. Lose not an hour. Those flabby muscles, bony cheeks, sunken eyes, thinned neck, ghastly looks, abundant stools, &c., are ominous of ill events. A very few weeks will effect a cure, otherwise it may be

" As the snowflake on the river,
A moment white, then gone forever ! "

With freight trains, the necessary regularity of habits cannot be enjoyed. They travel mostly at night, when the air is heavy and nature dictates repose. They push too rapidly along to be of service to us, and for the interests of trade, &c., stop in irrigated or dusty districts. Often they travel in unsuitable directions, do not go through the whole distance, are mostly Mexican, and speak only that language; do not stop at the most healthful camping places— where invalid trains should lie over for days or weeks—or places of grand scenery. At many of these there are hot springs, which may be advantageously used, with proper discretion, even in consumption, when a rheumatic habit has been discovered.

These, as most frontier countries, have immense stock raising interests. Most towns and ranches have large corrals for stock, are situated near streams which in all countries give a more or less humid air. So proverbially healthy are these parts that little attention has been given to the removal of manures, &c. The earth many feet deep is charged with absorbed animal matters which constantly give a noxious effluvia which patients should avoid. Only in well conducted invalid trains can the fullest advantages of these climates be enjoyed.

Whilst the susceptibility of the system to new impressions is strong the best use of them should be made. By neglecting this many have been merely *improved* who, in a special train, would have been *cured.*

Dr. E. Andrews, of Chicago, has recently compiled the following statistics from the U. S. census of mortality :

" The last two census reports (1860 and 1870) issued by our govern-
" ment have each a quarto volume showing the number of deaths in every
" State and Territory and the diseases causing them. By classifying the facts

"there stated it appears that consumption and cancer are two diseases which
"are similarly affected by, and prevail in, the same regions. The laws gov-
"erning their prevalence are two in number:

"*First*—These two diseases are most abundant near the sea and diminish
"as you recede from it.

"*Second*—At equal distances from the sea they prevail most at the north
"and diminish as you go south.

"For example, if you begin at Massachusetts and go westward, the
"proportion of deaths from consumption to deaths from all causes regularly
"diminishes as you recede from the Atlantic. Here are the figures:
"Deaths from Consumption in Massachusetts, 25 per cent; New York, 20
"per cent; Ohio, 16 per cent; Indiana, 14 per cent; Illinois, 11 per cent;
"Missouri, 9 per cent; Kansas, 8 per cent; Colorado, 8 per cent; Utah, 6
"per cent; and then, if you go down to California, it increases again to 14
"per cent, on account of the proximity of the Pacific Ocean.

"A similar decrease is observed if you go from north to south, as follows:
"Michigan, 16 per cent; Indiana, 14 per cent; Kentucky, 14 per cent;
"Tennessee, 12 per cent; Alabama, 6 per cent.

"From this, it follows that the best resort for a consumptive or cancer
"patient is some point which is at the same time as far south, and as far from
"the sea, as possible. Such a place is New Mexico, where the deaths from
"Consumption are only 3 per cent; while in cold and sea-girt New England
"they are 25 per cent. Probably the uplands of old Mexico would be still
"better."

We would remark that much of this mortality in Colorado from consumption, and some of that in New Mexico, is of persons who went there too late.
The mortality of Colorado and New Mexico from this cause, among the citizens who were healthy when they went there, could not be much greater than
that of Mexico, which is less than one per cent. Among those affected when
they there arrive, if the mortality were as great in proportion to numbers thus
diseased as in Massachusetts, then would Colorado show more mortality from
this cause than does that State. Alabama, unknown as a resort for such
invalids, shows only the mortality among the permanent citizens from that
cause.

Persons from abroad, by the water, air, food, &c., of localities where consumption is common, improve under the new influences; then authorities
attempt an explanation, by vague opinions, that because of temperament, or
some peculiar condition of the constitution at certain stages of the disease,
&c., conditions which they have totally failed to explain, the climate suits
certain classes of cases. As well might they say that the well water of Louis-
ville, Ky., purged newcomers because of temperament, idiosyncracy, &c., and
that if they remained long then the *temperament* changed so as to protect

them from the effect of that water which is borne with impunity by the permanent citizens. The only rational explanation is the *loss of susceptibility by continued use.*

Can it be possible that consumption is so protean in shape as to require so many directly opposite conditions for its cure? If so its proper management would forever have remained an unraveled mystery!

Not so, however; for those this day afflicted, where there is anything left to build upon—and in numerous instances very little appears to have been sufficient—any one thus diseased can satisfy himself or herself, in a few days or a very few weeks, that the general course to be pursued is alike suitable for all.

The fact that not only consumption but almost every form of chronic disease, of both sexes and all ages, bears a proportional ratio to our approach to the opposite of an active life in the free air of mountainous and other healthy localities is too true to ever grow stale, or the worse of wear from repetition by all authority on the momentous subject. The effeminate teacher or pupil is not worn down by mental labor, as generally believed, but by sedentary life in a vitiated and pent-up atmosphere. Mental labor, in connection with physical, is borne with entire impunity, and far greater strides made in literary advance. It is positively true that sedentary life in a humid and impure air is an abundant source of scorbutic diseases, and unmistakably indicates the actual necessity of a total revolution of habits, not only to prevent, but to cure disease whilst within the pale of the most certain and, in fact, only perfectly reliable means.

By reference to authority we find many writers who say that pulmonary invalids are improved by a sojourn in certain localities, but that too long a stay at such places is followed by a reverse effect. None, however, appear to attribute this loss of improvement to the real cause—loss of the new influence, although the necessity of change, of more than one change, of frequent change, until perfectly cured, would appear indicated by advice to these invalids to go abroad.* Many aver that foreigners improve in the West Indies and the Brazils, even in the parts where a large proportion of the natives die of consumption, but that " they cannot bear a long residence there without injury." Who can be surprised at this or fail to understand this two-fold effect? The correct explanation is, that the same morbific influences which affect the natives act also on newcomers, but the advantage of *change* for the while more than counterbalances the deleterious result, until the power of the new excitement is lost; then the fixed influence gets the mastery. For humanity's sake, and for the sake of *truth,* vital *truth,* we must urge such invalids to try a succession of changes, and if they must have a local sojourn, a permanent residence, that they seek it where a "large proportion" of the natives do not die of a disease they wish to avoid.

*For any digression from received opinions we beg indulgence on the ground that a wide range of discrepancy is shown amongst authors of equal eminence, who are respected even for conflicting opinions.

No mass of evidence can confute the statistics of Mexico. Could Cuba, to which so many consumptives have been sent, have had the climate of that country, the curability of that disease long ago would have been a fact established beyond all plausible controversy.

Persons going constantly in the open air of country localities, *even when exposed to sudden and great vicissitudes of weather*, are so generally exempt from pulmonic diseases that eminent authors, mistaking the power of the one influence alone, have intimated the necessity of exposure to sudden atmospheric changes as a preventive and curative, notwithstanding physicians of equal distinction, and, in fact, the majority of the profession, insist upon an equable temperature, constantly maintained. A solution of this discrepancy may be readily seen in full view of the immense power of exertion in the free air, which more than counterbalances the untoward circumstances of sudden changes. The fact is positively this, if a great majority of these invalids may be cured, when taken in time, by out-door life, without regard to conditions and vicissitudes of weather, a still greater number can more certainly be cured by the same exposure in better climates.

Cold throws upon the lungs humors productive of tubercle, causing successive and fatal crops of those bodies. Cold cannot be avoided the necessary length of time except by cautiously conducted out-of-doors life, in a suitable climate.

Humidity being recognized as the first amongst the predisposing causes of pulmonary disease, it is unaccountably strange that the almost anhydrous atmosphere secured by migrating with the changes of seasons, in suitable climates, was not earlier seen as the sure means of prevention and cure. And especially that such persons were sent to the humid, hot, and relaxing climate of southern latitudes, where relaxation promotes absorption and consumption of all the tissues, as shown in the West Indies, where the disease is very prevalent and fatal.

Our proposed course is the only mode of treatment which has given universal satisfaction. From the immense numbers we have met, even those who went too late, we have heard no complaint of the climate, except of those who remained too far North during winter and early spring; no expression of intention to "change doctors." An eminent physician from Pennsylvania said to me: "If I had only come to this climate sooner, instead of taking medicines, it would have been so much better for me."

Consumption increases in frequence of occurrence and in mortality with advancing civilization, showing the necessity of habits of men in a primitive condition to prevent, palliate and cure it. America offers the finest of all climates for this purpose. Dr. Rush associated the disease with civilization, for he remarked it was unknown to the North American Indians before they associated with white men, and was not found among colonists in the earliest stage of civilization. This would have been better worded if he had said of the Indians, before they partook of the luxuries, narcotics, &c., of the white man, and, to some extent, imitated his habits and emulated his example.

Henry W. Fuller, M. D., Fellow of the Royal College of Physicians, London, &c., &c., said:

"Of late years it has been asserted not only that remissions of the disease (consumption) may occur, but that the malady may be permanently arrested. In former days, such an event was regarded as simply impossible, the very fact of recovery being admitted as conclusive evidence that the diagnosis was at fault, and that the disease under which the patient was suffering could not have been phthisis.

"Even at the present time, there are not wanting those who maintain and act as if they believed that consumption is incurable. But there are no valid grounds for such an opinion. There is no reason why the constitutional derangement, on which the formation of tubercles depends, should not be checked, and the tubercles already formed got rid of. If the amount of tubercle already deposited be not excessive, and the general health can be improved, and the tubercular cachexia got rid of during the intervals between the attacks, the tendency to the further deposit of tubercle will be prevented, and virtually the patient will be cured. The tubercle already deposited may be absorbed, or it may remain unchanged and imbedded in the lung, or it may be softened and got rid of by expectoration; but in either case, if no further deposit of tubercles takes place, the functions of life will not be materially interfered with, and the patient may attain to longevity."

The following, from Dr. Fuller on consumption, needs a little comment.

"A still more prevalent, and a most mischievous error relates to the pre-"disposing agency of cold and wet, and atmospheric vicissitudes. Nothing "is more common than to hear consumption attributed to the effect of wet "and cold, the inclemency of the season, or the variableness of the climate "in which the person is resident; and nothing, unhappily, is more consistent "than the practice based on this supposition. Incased in warm clothes, "and shut up in heated rooms from which every breath of fresh air is "carefully excluded, the unfortunate victim to this prevalent superstition "becomes rapidly enervated and falls an easy prey to his disorder. But "careful observation, which is amply corroborated by statistical records, "proves incontestably that the pure air of heaven which God has provided "for us to breathe, and the variations of temperature to which in His all-wise "providence He has seen fit to subject us, are not so noxious or productive "of ill health as man in his ignorance has oftentimes asserted. * ".. * * The confined atmosphere of the dwellings, the want of sufficient "active out-door exercise, the various depressing passions, and the exhaustion "of mind and body resulting from the anxious struggles and vicissitudes of

"life, have far more influence than wet and cold in preparing the way for the
"inroad of consumption. Indeed, those classes of the community who lead an
"active out-door life, though necessarily exposed to wet and cold, are precisely
"those who are least subject to consumption. Even when the disease is
"already developed or far advanced, the pernicious effects resulting from over-
"caution in respect to exposure to atmospheric vicissitudes are often painfully
"apparent. Nowhere is this more strikingly exemplified than at our large
"hospitals, where the consumptive out-door patient who, ill clad and often ill
"fed, in hot weather and in cold, on wet days as on dry, has come to the hos-
"pital twice a week for medicine, notoriously lives far longer than his brother
"who, more fortunate in being well fed and protected from the inclemency of
"the weather, is shut up in the equably heated wards of the hospital, and thus
"loses the bracing, invigorating stimulus of the fresh breezes of heaven."

Any observing and thinking person may readily see that this by no means
shows that wet and cold are not producing causes of this disease, but that
they are only the lesser of two evils, as compared with the opposite condition
which he describes. Dr. Fuller could not undertake to say, with any plausi-
bility whatever, that similar out-door life and exercise in climates where wet
and cold are almost unintelligible terms, would not be far better.

Whilst in many densely populated parts of Europe and America, one-fifth
to one-eighth of the whole population die of consumption, showing a mortality in many places of not less than twenty per cent. from that disease,
so far as we can learn the Republic of Mexico shows a mortality from the
same cause of only seven-tenths of one per cent.!

To sum up, then, the fortunate results anticipated from our journey are to
be the offspring of a rare combination of favorable circumstances, embracing
attention to the condition of the mind, food, drink, air, exercise, bathing,
bedding, clothing, expansion of the chest, changes of place, with change of
atmosphere, diminishing pressure on the lungs during winter, and to the medicinal correction of any indisposition which may, from time to time, occur, by
one whose studied vital interest it has been to know these things, and who
feels that gratitude to his Savior, and duty to his fellow man, require that his
whole mind and heart should be dedicated to this cause, until the fact is established that, in like manner to the finding of the builder's rejected stone, so the
sanitarium for all invalids has been discovered, and that many who appear to
be hopelessly afflicted may be restored to relatives and friends, and to a life of
usefulness.

We most need the co-operation of those who can influence their slowly
but surely dying friends. Lose not a day. Procrastination is worse than a
thief of time; it is a thief of life. Dr. Edwards, a gentleman of great literary
culture, and an eminent physician, wrote from Chihuahua, that for him it was
too late: that if he had not tarried so long at San Antonio, although a fine

climate there for lung disease, he could have saved his life. He hoped this information would be the means of saving others.

Nothing earthly could afford us the happiness which we hope to realize from this enterprise in humanity's cause.

Nowhere have we found anything in print giving more in the same number of words than this:

"We have seen that the causes which are most instrumental in inducing "tubercular cachexy by suppressing or lessening the excretory and absorbent "functions of the skin, and in vitiating the blood, are humidity, a sedentary "mode of life, and the depressing passions. Now, the principal advantage of "a mild, dry and sunny climate in winter, is, that it places patients in the "most favorable conditions for counteracting the influence of these causes, "enabling them to take daily exercise in the open air, by which the muscu-"lar, respiratory, digestive and cutaneous systems are maintained in healthy "activity; whereas, in a cold and damp climate, such persons must necessarily "pass many days within doors, breathing the close atmosphere of warmed "rooms, and must moreover be deprived of the mental diversion which is "afforded by the variety of objects met with in walking or riding. Thus in "any such climate the quality of the blood becomes improved, and the ten-"dency to the formation of tubercle is diminished. The nervous and muscu-"lar systems experience the beneficial effect of this amelioration, which is "manifested by an increase of tone and vigor. The *moral* is likewise agreea-"bly impressed by the contrast which sunshine in winter presents to the cloudy "and rainy skies of which a lively recollection is retained. 'Is it not true,' "asks M. Foissac, in referring to the effects of light on the disposition of the "mind, that in bad weather the mind is more disposed to melancholy? Is "not British spleen occasioned, or at least kept up, by the thick fogs which "constitute for the inhabitants of Great Britain an atmosphere of dullness "and ennui? Are not petulency and vivacity excited by the aspect of clear "skies and sunshine? That they are so, is proved by the animated gestures "and the expressive play of features of the natives of Southern climes."

And from the same work, (Lee's dissertation to which a prize fund was awarded in 1856), we get the following: "'When the equilibrium of the "'moral acts is destroyed,' remarked a late medical writer—no less distin-"guished on account of his learning than for the elegance of his style—"we "may be sure that that of the vital actions will speedily be so. The physician "Elie maintained that four-fifths of men die from grief; an assertion much less "paradoxical than might be supposed; for truly, there are few diseases which, "in the actual state of our civilization, are not the reflex action of some strong "moral affection. It is the certain result within a given time, which must be "measured according to the violence of the attack and the individual disposi-"tion. Aneurism, liver enlargement, scirrhus, softening of the brain, most

"nervous diseases proceed more or less directly from some misfortune, experi-
"enced, it may be, long before, but of which the weight, the remembrance
"have at once broken down, or gradually weakened the springs of life. No
"one, therefore, dies of grief, of despair, nor of lost illusions : it is gastritis,
"pericarditis, apoplexy, which take the place, by their evident effect, of the
"real and active, though hidden principle of so many evils. Acute and pro-
"found moral suffering is, then, the point of departure of the greater number
"of organic alterations."

PREPARATION FOR THE TRIP.

Properly fitted up for this journey, persons may live as economically as at
home. Fish, fowl, deer, elk, antelope, sheep, buffalo, bear, turkey and other
game is abundant. Grasses in great variety exist in vast profusion. Horses
on slow trips do well on it, without corn.

During the winter of 1870–71, corn at Chihuahua and in many parts of
Mexico, went begging for bidders at sixty to eighty-seven and a half cents a
fanega, which is a measure of a little over two and a half bushels. Flour and
assorted meats are abundant.

A spring wagon, ambulance, or some kind of a long and straight-bodied
vehicle must be procured. At St. Louis these may be selected in great
variety with harness, at from $100 to $250. The bed should be long enough
for a mattrass, and to be used as a bed night and day, and if shallow and wide,
so much the better for ventilation, &c. They can be readily fitted up for
either one or two horses. Any of these wagons, if not found already so fin-
ished, may readily be fitted up with bows or some kind of frame-work to support a cover or top and side curtain to roll and button, hook or pin up. These
side curtains are indispensable to the necessary ventilation and to the free
admission of the balmy air when the weather is propitious, as it generally will
be. For the covering, a good article of white cotton duck will do as well as
the different articles of patent leathers, &c. A very neat little light wagon is
now offered, with springs fitted on the sand bolsters between the standards.
These are movable, so that when done with the wagon may be well sold for
either a light farm or spring wagon. The springs should be soft, and no
freight taken in these carriages, except the mattrass, camp-chair or seat, lunch
basket, &c. Arrangements, at very low figures, for the conveyance of all
necessary supplies, cooking utensils, &c., will be easily effected. Freight is
often taken five hundred miles for three cents per pound ! The roads generally through these Territories and Mexico are very excellent, teams passing
over them so easily that many doubt the distance between points. After
meeting, organizations for the conveyance of a cooking-stove fixed on a cart,
ready to be heated up in a few minutes after arriving at camp ; the services of
a cook, &c., may be readily obtained. Everything necessary to make the

expedition a pleasing recreation should be attended to, but much less arrangement will be necessary than at first will occur to persons inexperienced in camp life. A tent in this climate, although not particularly objectionable, is not at all necessary. Many who start with tents either dispose of them or store them away in the baggage wagon. Instead of an atmosphere pent up by the walls of tents, we want vivifying light, and air "quite unbreathed before." To increase the shade capacity of an ambulance at camp, the curtains may be drawn from a perpendicular, by cords through the lower button-holes, and tied to pegs fastened in the ground some distance from the wheels on the sunny side. For dressing of ladies and others the ambulance may be very conveniently used. For bathing, &c., each company should have a tent suitably situated. Tents taken for erection at places of two or three weeks' sojourn should be so constructed as to admit of free ventilation.

Much clothing and bedding is not necessary. A light and soft wool and hair or moss mattrass, one light and two heavy blankets, a shawl or overcoat two or three changes of light worsted clothing, a few hickory shirts and worsted undershirts, and a shotgun or repeater well nigh constitute this outfit.

All necessary supplies may be bought along the route. For any medicines or medical attention required, the customary charges will be made.

The whole length of our journey south of Santa Fe until we reach South Mexico, is infested by Indians and Indianized Mexicans and Americans. Entire safety consists not in action, but by preparation for action. They are not actuated by malice, but by stealth, and, therefore, are not dangerous when any risks of their lives are to be taken. Even small parties well prepared for them are never attacked, while the unwary are often molested. Through these parts, stock must be guarded night and day. Two or three well-armed men are equal to a score of these marauders. Mexicans, well recommended and reliable, may be engaged for a mere bagatelle, but a mixed guard is the best. Traveling parties have no trouble in effecting all necessary arrangements. Gentlemen attendants along with the party can arrange these watches among themselves, and accept compensation from those who need their services. Those who can afford it should have an extra horse or saddle pony. We shall earnestly urge them to live as much as advisable in the saddle. A very advantageous arrangement, readily effected after meeting, is that of two traveling together, one having a harness, horse and wagon, the other a saddle horse. Many roaring mountain torrents, picturesque lakes and towering summits, which direct attention from nature to nature's God, are accessible only on horseback.

Every article of outfit for this expedition may be well disposed of in Mexico by those who will have sufficiently recovered to return by southern lines; and those returning with us may exchange good or fine American horses or mares or large mules for a great number of the native animals, work and drive them back, and realize large profits, as well as the salutary influence of pleasing diversion.

Persons wishing to go are requested to so inform us until July 1 at Omaha, to July 12, at North Platte City, Neb., and July 20, at Greeley, Col., and after that all correspondence should be addressed to Denver, Col. This will enable us to give them much important information. It is especially desirable that we meet at Omaha by July 1.

If desired, everything except the horse and wagon will be furnished at moderate rates. But all, except a limited few, must have a horse and wagon, or at least a saddle horse or pony. These can be purchased anywhere along the line of our travel. It is especially desirable that invalids wait not for preparation, but immediately join us. Everything in our power shall be done to make them comfortable.

Although very desirable that patients should start at once, and join us immediately, the sooner the better, those who cannot do so may meet us at Colorado City in September, on our way southward from the parks, whilst the more tardy may, by correspondence with us, learn how to meet the train in South Mexico at mid-winter and enjoy the advantages and grandeur of the mountains and parks on our return next summer.

www.ingramcontent.com/pod-product-compliance
Lightning Source LLC
Chambersburg PA
CBHW021536270326
41930CB00008B/1273